Gone for Good

Gone for Good

*Tales of University Life
after the Golden Age*

Stuart Rojstaczer

OXFORD
UNIVERSITY PRESS
1999

OXFORD
UNIVERSITY PRESS

Oxford   New York
Athens   Auckland   Bangkok   Bogotá   Buenos Aires   Calcutta
Cape Town   Chennai   Dar es Salaam   Delhi   Florence
Hong Kong   Istanbul   Karachi   Kuala Lumpur   Madrid
Melbourne   Mexico City   Mumbai   Nairobi   Paris   São Paulo
Singapore   Taipei   Tokyo   Toronto   Warsaw

and associated companies in

Berlin   Ibadan

Copyright © 1999 by Oxford University Press, Inc.

Published by Oxford University Press, Inc.
198 Madison Avenue, New York, New York 10016

Library of Congress Cataloging-in-Publication Data
Rojstaczer, Stuart.
    Gone for good : tales of university life after the golden age /
by Stuart Rojstaczer.
        p. cm.
    ISBN 0-19-512682-3 (alk. paper)
    1. Education, Higher—United states.   2. College teaching—
United States.   I. Title.
LA227.4.R65   1999
378.73—dc21   98-45609

9 8 7 6 5 4 3 2 1

Printed in the United States of America
on acid-free paper

To Holly and Claire

CONTENTS

Preface   ix

1 Introduction
   why no one seems to know, even my mother,
   what I do at work   1

SECTION ONE Undergraduate Life

2 Lowering the Bar
   why we have such low intellectual expectations for students
   even though they could easily do more   13

3 The Prestige Business
   what services the university provides students and why we
   charge so much for tuition   27

4 Shortening the Yellow Brick Road
   why we have made college easier, yet no one seems to
   mind or care   37

5 The Sports Machine
   how universities entertain their students and alumni and
   why and how we've crossed the line of good judgment   47

SECTION TWO Research and Graduate Education

6 Heart and Soul
   why graduate students are often more important
   than professors   61

7 Grants or Goodbye
   why we spend so much time writing our grant proposals   72

8 Why Research?
   what professors do when they don't teach and why
   they do it   84

SECTION THREE Campus Politics

9 Matchmaking
   how we hire and why we move to other universities   95

10   The End of the Golden Age
     *why the era of exponential growth has ended and why
     it's a good thing that it's over*   106

11   Shaking the Tree
     *why universities are increasingly turning to alumni,
     foundations and corporations, and what they will and
     will not do in exchange for money*   119

12   You've Got to Believe
     *why we blindly follow the latest trends in academic
     fashion even though it makes us look ridiculous*   129

13   The Fifty Percent Solution
     *why there are so few female professors, and why there
     aren't likely to be more in the foreseeable future*   140

14   Making Adjustments
     *how to adapt to the life of a professor without getting
     too crazy*   152

15   Getting Tenure
     *what it takes to get tenure, why standards have risen,
     and why they will continue to rise*   163

16   Rolling the Dice
     *why the American university is still valuable even
     though it looks to be in such a mess*   178

Preface

When I originally began to work on this book, I had recently received tenure and I had in mind writing a light, humorous guide called something like "How to Get Tenure." A friend of mine, John Cassidy, writes humorous books for children and adults, and I was thinking of something along the line of his Klutz Press books. Perhaps, following the Klutz recipe, I would include, attached in a net bag, a "useful" item for a professor, like a pipe, or leather elbow pads for sewing onto a tweed jacket. I was going to collect friends' and my own humorous experiences as untenured professors, include a board game called "Publish or Perish," and write a multiple-choice quiz that allowed readers to evaluate whether they were tenurable material.

But as I wrote, I found that much of the humor was being crowded out by serious issues that kept popping up, and I had to change course. I had spent seven years working during a very difficult transition time for universities that still has not ended. While there have been many humorous incidents along the way, the overall backdrop has been one of high levels of discontent. Both professors and administrators have been disenchanted with how university life is changing and the press, the government, and the public have been disenchanted with how out of touch universities are with the needs of society. The two sides, those inside the university and those outside the university, seem to be so far apart that they are unable to communicate. So, instead of writing a laugh-a-minute book about a rather unhumorous time in university life, I decided to write a personal chronicle of this time period. When I remembered a syndicated op-ed piece on funding trends for research universities written by the Cal Tech physicist David Goodstein, I knew that I had found the theme for my book.

It should be noted that some of the locations, dates, and personal descriptions of the events and conversations described in this book have been altered to obscure identities. In my view, all students, faculty, and staff in a university are part owners and I frequently refer to Duke as "my university." This book contains both praise and criticism and it should be understood that neither element is directed at Duke in particular. I consider Duke University to be fairly typical of universities across the nation in terms of its governance, goals, and aspirations. Had I been at another university, I would have written much the same kind of book.

I would like to thank Peter Haff, a geologist colleague of mine, for being a sounding board for many of the ideas and stories in the book during our lunches on "The Wall." The deans and other senior administrators at Duke, whose doors were generally open to me (at least until they found out that I was writing a book on university life), were valuable adversaries with whom I had many discussions that helped clarify my opinions. My post-Golden Age friends at other universities have generously allowed me to include their stories in this book. For obvious reasons they remain anonymous.

Much of this book was written while I was directing an overseas undergraduate program in Italy. The staff at Venice International University provided logistical help, a pleasant office, and good cheer. My original contract with Oxford University Press was for a hydrology textbook and Joyce Berry deftly made the transition to editing this book instead. I hope that it is a worthy substitute. Who knows? I may still write a hydrology textbook. David Bell, Andrew Burness, Barbara Gaal, George Hornberger, Steven Ingebritsen, Michael Lavine, Gustavo Perez-Firmat, James Roberts, and Gloria Welstein kindly took the time to provide helpful critiques of draft versions of this book. Thomas Harkins at the Duke Archives and the reference librarians at Perkins Library tracked down numerous facts. The office of Congressman F. James Sensenbrenner provided information on federal funding of universities.

My mother spurred me on with the following weekly question during our phone conversations: "So Mr. Writer, is the book finished yet?" I would especially like to thank my wife, Holly, and daughter, Claire. My wife served as my first reader and her detailed editing vastly improved the manuscript. My daughter edited and provided advice on methods of writing that she has gleaned from her favorite authors. This book would not have been completed without their constant encouragement.

Chapel Hill, North Carolina   Stuart Rojstaczer
December 1, 1998

Gone for Good

1

Introduction

It is the second week of August 1990. The U.S. government is determining how to respond to Saddam Hussein's invasion of Kuwait. On a less serious front, the Oakland Athletics, my beloved A's, are comfortably in first place in the American League Western Division and are playing with confidence. But I am far from Oakland, having just arrived at the airport of my new address, Durham, North Carolina, where I have taken a position as a not-so-young assistant professor of geology at Duke University. I have been here twice before, once for my job interview and the second time to buy a house with my family. School will start in a week, my wife and six-year-old daughter will arrive in a few days from my in-laws' house in Chicago, and students, mostly from Duke, are filling the airport on the way to their dormitories. For me, having lived in much larger cities almost all of my life, the airport seems like a miniature version of the real thing.

I look at all of the students in the airport who, like me, are waiting for their luggage to arrive. I have spent the last several years working in a government research lab where almost everyone was at least a decade older than I. The era of the well-funded government research lab was coming to a close, so I decided to find a job as a university professor. After those years in the government, the sight of all of these young people is foreign. Bernard Baruch, the financier, once said "Old age is fifteen years older than I currently am." If this is a universal truth, then I must look old to many of these students. Conversely, many of them look to me like they should still be in high school. The male students tend to be long of limb and absent of embellishing muscles. The female students tend to be slimmer and their voices are a little more high-pitched than I am used to. They are all college bound, some of them will graduate this coming year, and they are my audience for as long as I keep or am allowed to keep my position (tenure is granted to about 40 percent of those who enter as assistant professors at Duke University).

I think about what it will be like to teach these students. They seem to be serious—there are minimal amounts of laughter even though they form a critical mass—but maybe they are just nervous and preoccupied. I know that, according to test scores, they are close to the upper echelon of students in this country. Several years before 1990, *U.S. News and World Report* began rating the relative merits of colleges much like the Associated Press rates college football teams. By 1990, Duke had established itself in the top ten in these rankings. Despite the dubious value of this type of evaluation, many students and parents of students make use of it in their selection of colleges and, as a result, the quality of the applicants and number of applicants at my university have risen with its ranking.

The college guides that I browsed through in the Palo Alto, California, bookstore that was within walking distance from my home suggested that Duke is a place for the smarter children of suburban, upper middle-class and upper-class families to spend their undergraduate years before they go on to professional school. This profile didn't sound a whole lot different from most expensive, selective liberal arts colleges and universities. At the time, I guessed that these students were almost on a par with Stanford (where I received my Ph.D.) students, but given their eastern and southern upbringing, were somewhat less adventurous.

I will be reminded of my preconception of our students at the end of my first school year when I will, for obscure reasons, be in Philadelphia looking for a hotel in which to spend a night. Wearing a Duke University T-shirt I will walk into a dumpy hotel and ask the desk clerk for the daily rate. He will quote me a rate of sixty-five dollars. I will pause for about ten seconds and he will look at my shirt and say, "Look, Mr. Duke University, you ain't gonna stay here. A Motel 6 would be cleaner and cheaper. If you were from Harvard you'd already be out the door."

Whatever preconceptions I have in the airport about these students and teaching, however, are probably meaningless. I have no reliable data with which to make an evaluation and as a scientist I know that I am on shaky ground. I had done very little teaching, just one semester at the University of Illinois. I have had very little exposure to undergraduates at private liberal arts universities and no exposure to a university that primarily draws its undergraduates from the East and South. I am so green about this process that a student or parent of a student may wonder why I was hired. The simple answer is that I wasn't hired because of my talent and experience as a teacher. I was hired because of my potential as a researcher, and the hope is that I will take to teaching as well. Such have been the hiring priorities of most universities since the 1960s.

People have told me what being a professor at a liberal arts university will be about, but taking a job like this is a little like having a baby. Any advice seems very abstract. Like having a baby, all you really know is that it will change your life and there is little you can do to prepare except get as much sleep and do as much socializing as possible beforehand.

The airport is about midway between Raleigh and Durham (ten to fifteen miles from either city), and there is no public transportation from the airport. I decide to save some money and take a shuttle bus from the airport to the campus where I can easily walk the few blocks to my house with my suitcase in hand. The shuttle bus is loaded with students and I get in a conversation with one of them, a freshman from Long Island, New York. He is a skinny, short young man with short black hair, angular features, and an earnest way of talking. I don't know his career aspirations, but he seems impressive. We're both new here, this is probably his first encounter with a professor, and this is definitely my first real encounter with an undergraduate student.

"How long have you been preparing for teaching your classes?" he asks.

"Well I teach one class this semester. I'm just getting started," I say, trying to avoid the question.

"Yeah, but haven't you been working on your lectures?"

"No. I've been working at my previous job."

The young man looks disappointed. His parents are paying $20,000 per year to send him here, and they've probably taken out God knows how many dollars in student loans to have him come and be instructed by some of the best (purportedly, at any rate) professors in the country. And now he finds out that at least some of his professors will be teaching on the fly. "Duke doesn't start paying me until September," I say. "You can't expect me to quit my job and work for nothing preparing class material."

"They don't pay you to get ready for teaching?"

"No. As a matter of fact, they don't even pay me for the first week. I'm teaching pro bono." The young man doesn't see the humor in this and as a matter of fact, when I first saw my contract, I didn't see the humor in it either. When the shuttle bus reaches the campus and we are all let off, I say good-bye to him and wish him the best of luck. I shake his hand and feel bad that I have crushed an illusion about university life even before he has officially arrived.

It's three years later on a Sunday in August. Saddam Hussein has long since surrendered occupation of Kuwait, but continues to confound UN inspectors of his weapons. Bosnia is being torn apart and virtually no one from outside is willing to try to help. My beloved Oakland A's minus their recently retired third baseman are floundering. But all of these issues, both major and minor, are not concerning me at this moment. I'm in my office and analyzing data that I have collected in the field. I'm a hydrologist—someone who studies rivers and ground-water—and I spend a fair amount of time making measurements in streams and in wells. The summer is when I get much of my data analysis and research done in a relatively uninterrupted fashion. There are no department meetings and university business to distract me. Undergraduates are gone and all that's left are faculty and the graduate students, most of whom in my department do research out of town during the summer (they are geologists and marine scientists). It's when I can concentrate the best and work on my most difficult material.

I know that it's fairly nice outside so at about two o'clock I decide to take a break. I leave my office and sit on the steps of my building and eat some yogurt. It's sunny and cool for the South in August, about eighty-five degrees. The sky is a dusty Carolina blue from all of the moisture in the air, a color that still seems faded to me after my years staring at the sky in the western United States. I'm about at the midway point before my tenure decision and I'm reflecting on my progress.

I'm in a cocky mood. By all measures of a research university, I am doing pretty darn well. I'm getting significant amounts of grant money and I'm publishing in highly visible scientific journals. I'm starting to attract excellent graduate students to work with me. My classroom teaching of graduate students has been a positive experience for both the students (as they indicate in their end-of-class evaluations) and myself. My undergraduate teaching has produced only average student evaluations, but it's clear that one significant reason for this is that I am a tough grader and expect students to work harder than in most classes. My department chair (who is more or less my boss and is the single most important person with regard to my future tenure decision) and I, after a very rocky beginning, seem to be getting along reasonably well.

I've turned down a job offer from another university and in the unusual state of affairs in academia it's almost always a good idea to let your home institution know this. In many workplaces seeking new employment, getting outside offers, and using them (either on purpose or inadvertently) to gain leverage would be thought of as being disloyal and be a potential cause for being fired. In a marriage, this would be the equivalent of flirting with someone at a party with your spouse in full view, arranging to meet for a tryst at some hotel and never showing up. While this type of behavior would probably lead to divorce in a marriage and possible harm from your potential lover it is the norm in academia. Finally, my contract with Duke University, originally for four years, is being renewed. The letter I received announcing the renewal is filled with glowing assessments of my accomplishments and hints that if I keep up the pace of my research, I'll likely receive early tenure.

I'm looking out onto the verdant, manicured Duke University quadrangle, elegantly designed by a black architect from Philadelphia in the 1920s who (as the story goes) never visited the campus because he didn't want any part of the Jim Crow life of the South. A threesome is walking along the sidewalk. They are getting closer to me and I quickly recognize them as a prospective student and his parents. The young man, a rising senior in high school no doubt, is tall compared to his parents—about six feet two with pale skin. He must be a musician or a computer type, I think. It's August after all. His blond curly hair is short and his posture is poor with shoulders slumped. His head is sagging as if he is trying not to stand out in relation to his parents. We never participated in the "posture photos" of the Ivy League—a bizarre activity where photos of new college students without clothes were taken over a period of a few decades—but our students tend to stand erect and proud. So, at face value, I don't give this young man much chance of being accepted. The short parents—the father, rotund with a full head of blonde hair, and the mother, a brunette with a short perm who looks tired—are in their late forties to early fifties. They stand on either side of their son as if to buoy him.

The father walks up to me and asks me the name of the building in front of him. Most of the buildings at my university lack signs indicating their names. Perhaps this is a weak attempt to keep out the unwanted by making it hard for them to get around on campus. "Old Chemistry," I say. "You can tell from the limestone carvings of the test tubes and scales at the entrance." I turn toward the carvings and point to them. "They were carved by an Italian immigrant family who did all of the masonry. The Pellegrinis or something like that. It used to be called just Chemistry, but the chemistry department had a new building built for it in the sixties. Hence the name Old Chemistry."

The father looks at me quizzically. I have told him more than he wants to know. I'm still holding the yogurt container in my hand. "You're a professor here, right?" he asks. I nod. "Well it's summer and you don't have anything to do. Why don't you give us a tour of this place?"

I look at him. He's presuming that because classes aren't in session I must be twiddling my thumbs all day long. I'm not upset about this, though. Many people seem to have this view and I'm used to it. They assume that the sole "real" job of a professor is to teach or prepare to teach. They assume this because we have never made any real attempt to show the public what it is we do when we don't teach. They think that we spend our nonteaching hours and our summers accumulating more knowledge for teaching by reading books—if we're industrious. They also tend to think that we are lazy and unaccountable, and that many of us pretend that we are accumulating more knowledge for teaching and instead are just loafing around.

Let me put the family standing before me waiting for a campus tour on hold for a bit. I've talked to people on airplanes, I've talked to alumni, I've talked to neighbors, and I've talked to my mother, all of whom (even my mother, bless her heart) have had this false idea about what professors do when they don't teach. They don't know that we spend the equivalent of a full-time job doing work related to our research in addition to our teaching and administrative duties. If I am developing a brand new class, then I will spend another thirty hours a week doing work related to teaching as well. If not, then I typically spend fifteen hours or so doing work related to teaching (about ten hours) and university business (the remaining five hours) every week in addition to my research.

The student or paying parent will likely think that this ratio of work in research to work teaching is not fair. Tuition costs big-time dollars and they would like to get more of a professor's time for their money.

I basically agree with this request. But in a modern research university, the financial demands are high and the expectation is that you spend the equivalent of a full-time job doing research. We're expected to hold two jobs. By necessity, my teaching job, the one the parent is paying for, is part-time. Nowadays, I wish it were more balanced, but back then, sitting on the steps of the Old Chemistry Building as an untenured professor, I didn't even think about what was fair and what wasn't. However, both then and now, my feelings were that research is essential to any university.

From a practical standpoint, research is important because, at least in the sciences and engineering, research generates dollars. Personally, it generates a significant amount of my salary. Only 75 percent of my salary is guaranteed at my university. The other 25 percent must come from grants to perform my research. In essence, I work partly on commission. But the university benefits financially even more than I do from this activity. Grants associated with research are a major source of revenue for universities like mine. At my university, 40 percent of total university revenue for academic programs comes from research dollars.

Even ignoring money matters, research is important. For example, suppose you want to learn how to build a boat. You decide to take a class to do this. Who would you rather have as your instructor: someone who has read about building boats, someone who built one boat twenty years ago, or someone who has built lots of boats and is still building boats? At face value, you would say, I don't want to build a boat that might sink. Give me the teacher who has built lots of boats.

Well, the same can be said for university instruction. For example, take my field—hydrology—which includes the nature of lakes, rivers, groundwater and precipitation. Who would you rather have as an instructor: someone who has read a lot of books about it, someone who studied one river twenty years ago, or someone who actively studies rivers and lakes? At face value, the selection is a no brainer. You want the active researcher because that person probably really knows their stuff.

But I've gotten a little ahead of myself. Let's get back to the couple and their son standing before me on a pleasant summer day. They are intelligent people and they could easily understand what I do. But no one has ever bothered to tell them that part of what I do during the summer (forgetting about the contribution to science for a moment or two) is to ensure through my research that the university generates enough money so that tuition isn't even more than the ridiculous amount of money that it already is. No one has effectively informed them that through research a professor stays on top of his or her field and, as a result, can potentially teach a better class than the person who has never done research or last did research twenty years ago.

Should I give them a personal tour of the university as requested? On some level I am aware that if their son chooses to come here, I will benefit from the $100,000 they will spend for their son's education. But I don't really care about their $100,000. Far more people are willing to spend this kind of money on their children's education than we can possibly admit. The fact is we produce a product, though much maligned by the popular press, which is in great demand (The popular press seems to view university education in the same way it views the software maker Microsoft Corporation. They sometimes dislike our product, even though it's a product that many need and use.) So if I choose, I can just be arrogant (believe me, the American university isn't the only institution where arrogance is in far too ample supply) and politely or impolitely decline. I am just about done with my yogurt. If I want to get tenure, I need to keep plugging away on my research.

But being haughty with people I've never met before is just not my thing. They've probably come a long way and it is a Sunday after all. I also admire anyone who can be so direct (directness is not a common quantity in a university), so I say, "Sure, I'll give you a tour," and get up, leaving my yogurt behind a limestone pillar. While I show them the main quadrangle, they tell me part of their life story. The young man is the man's stepson, a child from the mother's first marriage. He's their only child so I feel some affinity with them as the parent of an only child. They're from upstate New York, where the father works as an electrical engineer. "I'm a Purdue man," the father says.

"Purdue. That's a strong engineering school," I say with the modicum of knowledge I've gleaned from teaching engineering students and taking engineering classes when I was a student.

"Yes, it is," he says. "I'd like George to go there, too. He can get a good education for a lot less money. But he likes this place."

"Well, we do have a very selective engineering school here, sir. And your son would receive a very personal touch. The class sizes are small. It's a very productive way to achieve a high quality education." I don't know why I'm laying it on so thick. I'm usually not a salesman in the least. I also feel that the chance is virtually nil, regardless of his high school record, that his son will come to my university. He's a Purdue man, I think. His son will be a Purdue man, plain and simple.

"That may be," he says. "But I received an excellent education at Purdue." He is none too happy with my attempt to sell him my university.

"I am sure you did. Wasn't there an astronaut who was an engineering alum of Purdue?" I say, remembering an advertisement I saw during the NCAA basketball tournament on television.

"There were several, but Neil Armstrong's the one you're probably thinking of," he says. "I was too young to overlap with him in school. A fine man."

"He was the first man to walk on the moon, right?"

"Apollo 11 was the mission he was on," he says, and I've returned them back to our starting place. We shake hands and say good-bye. I remind him of the quality education his son would receive at Duke.

The new students and their parents, the prospective students and their parents, I will meet them often. Their assumption about what I do and the workings of a university are generally so far off the mark that it's almost frightening that they are willing to spend so much money on something they know so little about. In my first seven years I will be reminded time and time again about the disparity between what parents and students want from a university and what they actually get as well as what faculty want from students and what they get.

There will also be a great disparity between my attitude as a neophyte professor and my attitude seven years later. The change in attitude undoubtedly reflected changes in my personal life. But mostly my change in attitude was initiated by fundamental changes in the economics of a university. For forty years, universities had grown dramatically in size and in number. Those forty years were the "Golden Age" of the American university. It was a period when growth was financed by Cold War-based government funding of research in the sciences and engineering, and rapid increases in tuition dollars from a growing student population that benefited from government-backed student loans. Universities were the source of major construction contracts and the size of their faculties, administrative staffs, and research programs mushroomed. When I entered, universities were optimistic that rapid growth in federal funding of research and education would continue unabated and ensure the growth of higher education nationwide. They felt secure that the Golden Age would continue.

But within a few years of my being a professor, that optimism had died. Universities nationwide were being hit with economic constraints caused by more or less flat revenue in the presence of rapidly rising fixed costs. It was clear that the Golden Age of the American university had ended.

The end of the growth years made me rethink my role as a university professor. I had entered as someone whose ideas about the relative roles of teaching and research in a university reflected the mainstream views of his profession. But at the end of seven years, my views and those generally held by my profession diverged markedly. While I still value my research and its role in the university, I no longer value it with the same near religious fervor. While I am still frustrated by the general lack of interest shown by undergraduates in learning, I no longer put the major burden of blame on the students. When I look at what has been fondly referred to as the Golden Age, I don't view it fondly anymore, but as a rather crazy time when the university's mission became too specialized for its own good.

I had entered as an untenured professor with limited teaching experience and a promising, but still untested research career. Over the span of seven years, I had been promoted and tenured and was both a much better teacher and researcher. And my views on the end of the Golden Age were much less bleak than those of many of my colleagues.

I believe that the end of rapid growth portends good news for the American research university. It means that the American research university has a reason to change and redefine its purpose so that it is more diverse and balanced. These changes, if they take place (and I believe that they will take place because historically universities have changed, albeit slowly, with the times), will create a university that is, on the whole, more beneficial to students and to society.

This book is a personal chronicle of university life. It is designed to give parents and students and anyone else who may be interested a window into how an American research university currently works and how it has more or less worked for the last fifty years. The book is also designed to show how the current model, the model of the Golden Age, is an inadequate one for today, was inappropriate (although economically attractive until recently) for the past two to three decades, and could be relatively easily changed for the better.

SECTION ONE
Undergraduate Life

2

Lowering the Bar

It was the first day of my first undergraduate class. The lecture hall had a capacity of about 150 and was a bit like a movie theater with a sloping floor that rose gradually with distance from the lectern. Although old, the room was well designed. A three-paneled chalkboard covered the front wall. Every seat was relatively close to the front of the room. There were ample windows that gave the room excellent lighting.

Sixty-five students sat in front of me ready to begin an introductory class in environmental science. They looked collectively nervous and shy, which is typical for a first day. But I wasn't nervous. I've been comfortable in front of crowds my entire adult life. I introduced myself. I'm aware that first impressions are very important and I tried to be light-hearted and energetic. I welcomed the students and then made a joke about my funny last name. I'm a bit goofy in front of a classroom, which usually helps bridge the gap between the students and myself. When I was a kid, there were two kinds of professors that I noticed in movies. There was the imposing imperious professor, à la John Houseman in the film *The Paper Chase*. Then there was the goofy one, à la Fred MacMurray in *The Absent Minded Professor*. I tended toward the absent-minded model, a kind of borscht belt version of Fred MacMurray. I hadn't taught in ten years, but my sense of humor still went over well with this generation of students. What I would learn, however, is that my expectations of student performance and approach to teaching science were no longer suitable.

I handed out the syllabus and laid out the structure of the class. I wasn't trying to overwhelm the students with a great deal of work. My objective was to create an attractive class that had high annual enrollments and enticed a few of the students to become majors in my department. I expected the class to study what I thought was a modest amount, about four to six hours a week. In terms of logistics, there were six homework assignments, one group project on an environmental topic of choice, and three exams. No late homework was allowed, but students were allowed to drop their lowest homework score. One of the students toward the front of the class raised her hand and asked, "Will the exams be multiple choice?" "No," I said. "They will be short essay exams." She made a face. I never saw her in class again.

I didn't know that my class structure was highly unconventional. Homework assignments, which I thought were essential, generally were not given in introductory science classes unless they were a part of the premedical or engineering curriculum. Exams with multiple-choice or fill-in-the-blank questions were the general rule for these introductory classes as well. Requiring students to write papers was not all that unusual, but requiring group projects that consisted of a jointly authored paper and a presentation was not done. My expectations were that they would think and work reasonably hard. They would feel good about the class because they would learn interesting material. Where did I get such ideas? Partly, I received them from my own education. I wanted to teach the way that I liked being taught.

As a student, the professors that I really learned from were the ones who stretched their students. Some did it with humor, and some with intimidation. Regardless of the exact method used, their major goal was intellectual discovery. They knew that most students would not go on to lead an intellectual life, but they wanted all students to be thoroughly exposed to the life of the mind while they were in college. This is the way I wanted to teach and it was the way I had taught very successfully, albeit briefly, ten years previous.

As a graduate student at the University of Illinois, I gave a weekly lecture and laboratory as part of an introductory environmental science class similar to the one I would teach at Duke. I liked the experience and it was one of the reasons that I decided to become a professor. Almost all of the students were nonscience majors and had a good work ethic. We developed a good rapport and I distinctly remember that the students felt positive about the material they learned. I also remember feeling satisfied with their level of achievement. The material I presented assumed that the students had remembered their high school math and with a little bit of review could use their high school chemistry. The homework was designed to give students a basic introduction to environmental problem solving. In a class of twenty students, I gave three As and A minuses, a lot of Bs and quite a few Cs and I didn't receive a complaint about grades.

Although I had been successful earlier, times had changed. My expectations of student performance, which were considered reasonable in 1980, were too hard nosed for the students of 1991. I was bound to scare students away from my classes. Much of the competition for students came from classes that were both easier and different in approach than mine. Ironically, the class that I had taught in 1980 represented, at the time, one of the easier ways to fulfill a science requirement. Classes in introductory chemistry and physics were much harder both in terms of workload and grading, and they still were in 1991. Over the intervening years, however, many universities had eliminated or weakened their science requirements. At Duke for example, a science requirement existed, but since 1986 it could be easily avoided, and over 10 percent of the students never enrolled in a science class. In response, professors who taught science classes that were not a part of the premedical or engineering curriculum felt compelled to make classes easier in order to maintain reasonable enrollments.

Also over the intervening time, a new wave of classes had taken hold in colleges and universities. In the social sciences and humanities, these classes focused on current events and popular culture. In the natural sciences, they focused on the broad achievements of science and abandoned the traditional assumption that in order to understand science, students had to learn by doing and solve problems. These "new wave" classes tended to be easier in terms of workload, intellectual expectation, and grading. They began to be introduced in the 1960s, and by the 1980s their number was large enough to constitute a college curriculum all its own. Students were free to mix or choose from two paths of study, the traditional and the new wave. Parents of students and conservative social critics have been more concerned about the overt leftist political advocacy of a small number of these new wave classes rather than their tendency to demand less of students.

I wasn't aware of this change in expectations and approach over the previous decade. And if I had known, I probably would have ignored it because I was confident that I could attract students with my own methods. I put the class together with loving care. I ransacked the library researching various topics. I thought long and hard about homework assignments and I included material that I thought was exciting and societally relevant. I gave the course an ambitious title, "Environment and Industrial Civilization."

Senior faculty watched me spending all this time on my course, and I felt I knew what they were thinking. If this boy spends this much time teaching there isn't a prayer that he will get tenure. But I ignored the hints that I received from senior faculty. Besides, I was doing research as well. I was putting in a good seventy to eighty hours a week to make sure that I got off to a good start. This meant, of course, that I was largely ignoring my family. Like many people, I was too caught up in my work to notice. In retrospect, I can't laugh about how caught up I was in my work, but I can laugh about how unrealistic my goals were.

During the second lecture, I handed out my first homework. I was a bit nervous about it. While it seemed a relatively easy tune-up piece requiring no more than about four hours of work, I really did not know what Duke students were capable of doing. As a check, I asked my wife to look at the homework and she thought it seemed reasonable. Then I gave it to my teaching assistant (who helped with grading the homework), who thought it was much too hard. Then again, this teaching assistant was a below average student. He was a Ph.D. candidate admitted from another country in exchange for allowing some of our faculty to do fieldwork in that country (politics are clearly a part of science just as they are in the rest of real life). So I didn't quite know what to expect.

The average grade for the first homework was 94 and I thought I was off to a good start. But the truth was that, even early on, the mood of the class was mixed. Sure, everyone could do the work I assigned. They were smart students, but many were expecting a light and easy overview that would fulfill Duke's optional science requirement. Instead I was giving them a reasonably thorough introduction to environmental problem solving. Ten years previous, my approach and expectations were part of the mainstream, but they weren't anymore.

So after a few weeks, when I looked out at the class, I saw three distinct groups. About 30 percent of them seemed to be enjoying the experience. Another 50 percent were of mixed emotions about the class. Some of them liked the material, but did not appreciate the workload. Others liked my teaching style, but did not like the material. Finally, another 20 percent were highly negative. This group either hated science or hated the amount of work or hated me. They, by and large, dutifully attended and took notes because they wanted to do well on the exams, but they were not enthused. A few were fearful of the prospect of taking an essay test in a science class. I saw the pain on some of their faces while I lectured.

I was not, however, like Bill Clinton. I did not feel their pain. I wasn't going to make the class easier to appease those who were disinterested. I wanted to give the motivated students their money's worth. So I focused on those who were really enjoying the class. In any given lecture (attendance was typically about 80 percent) there were about a dozen of them sprinkled throughout the lecture hall. When I asked questions of the class, they were the ones who participated in the give and take. As someone trained in the sciences, I was used to classes where the tendency was to gear the class to the scientifically talented. Mine was a kinder, gentler version of this approach. You didn't have to be talented in the sciences, but I lectured with the assumption that you were motivated to learn.

I mentioned to my colleagues that I was gearing my class to the motivated students. They were not supportive of this approach. But to my way of thinking at the time, what I was doing was the right thing. Nowadays, I still feel this way, but from a practical matter (my enrollments would be negligble if I followed my feelings), I have to spend more time appeasing the slackers.

I should say that I was (naively) surprised by the large number of slackers I had to teach in this class. I came to one of the better universities in the country and expected to teach bright, strongly motivated students. Most were very bright, but quite a few were not motivated. They were clearly here principally for the social aspect of college rather than for an education. Thinking about my own undergraduate years, there were also many students who were slackers. They went to school to earn their credentials so that they could go on in life. Educationally, they went through the motions, and focused on the parties and alcohol at night.

But I'm dwelling far too much on the poor students. There were many students in this class who were clearly interested in more than a grade. Having some slackers in this class was simply the price of teaching in a not so perfect world.

While not everyone liked my class, the students were almost always attentive and polite. Initially, I took this aspect of my class for granted. But then I gave a guest lecture in another introductory science class. The professor had not made much of an attempt to be serious about educating, and during my guest lecture the attention level was barely perceptible. People were reading the newspaper. They engaged in side conversations like I wasn't even there. I felt like a minor musician playing in a bar and not being listened to. I tried to get them to listen, but was unsuccessful. After the lecture, two students came up to the front of the class to talk to me.

"Professor R.," one of them said, "we want to apologize for the rudeness of this class."

"Is it always this bad, or is it just because I am a guest lecturer?" I asked.

"No. It's pretty much this way all of the time." After talking to them, I didn't feel so bad about having to lecture to this class. I felt much worse about these two students who wanted to learn something, but had to do so under compromising conditions.

My first class continued onward with the same mixture of enthusiasm and grimacing faces. The average homework grades continued to be high for all but the true slackers. The exam grades varied more widely. The average class grade on the short essay exams was about a B to B minus. I thought that this was pretty good for grades, but remember I had not taught for ten years. What was an acceptable average grade in 1980 was not acceptable in 1991 (or today). At my university, the average grade for students rose from 3.0 to 3.4 from 1986 through 1994 before it stabilized. Similar changes had taken place across the country in the 1970s to 1980s. In the eleven years since my first teaching experience, professors were grading easier almost everywhere.

There are at least a few reasons why grades have risen. One is that professors long ago figured out lowering expectations and giving high grades meant less time dealing with students and more time for research. In essence, there is an unspoken contract between the professor and the student. The professor agrees to provide an easy class in order to be left alone. The student "benefits" in having more time for social activities.

Second, it also doesn't hurt that easily graded classes attract more students. (I don't think students are behaving inappropriately by preferring classes with high grades. They would be foolish not to.) Departments with large student enrollments tend to get larger budgets and stand a much better chance of getting permission to hire new faculty. These incentives often drive professors to institute lax grading policies.

Third, the full emergence of new wave classes across the university has created a significant new source of grade inflation. The reasons why these classes tend to grade generously vary. Some of the professors of new wave classes find grades to be an antiquated standard, and that unbiased judgment of student performance is impossible. Most faculty do not agree with this "post-modern" view, but in order not to lose students to the new wave classes, they feel compelled to keep pace and raise their grades.

Probably a more subtle reason for the elevation in grades (at least at private universities) is the dramatic increase in tuition over the last twenty years. For example in 1976, tuition (including fees) at Duke, Harvard, and Chicago was $3330, $4090, and $3517 respectively. Over the following two decades tuition at most private universities rose at roughly double the rate of inflation. In 1997, tuition at the three institutions noted above was in the range of $21,000 to $23,000. When a student represents $22,000 in revenue, that student has a fair degree of economic clout. There is a tendency to want to reward that student in some way for choosing that university. One way of doing this is to give the student easy grades and an easy sense of achievement. It could be said that the student is inadvertently bribing the university. Also, parents like to receive good news from their children, and by inflating grades, universities assure that most students will be able to inform their parents that they are doing well in school.

Regardless of the causes, grade inflation has proven difficult to reverse. At Stanford since the mid-1990s, faculty have been trying to find a way to reduce their average GPA from a truly outlandish 3.6 to a modestly outlandish 3.4. Since my university's average GPA is already at the 3.4 level, I guess I should feel proud of our relatively stringent grading standards. Our attempts at reversing grade inflation, however, have not been successful. A 1997 effort at Duke to institute a rating of students that rewarded those who enrolled in difficult classes was defeated overwhelmingly by the faculty.

As is typical of elite private universities today, almost all of the departments at Duke give As and Bs to more than 90 percent of the students. Taken at face value, this indicates that the performance of students is almost always good to excellent. Relative to these standards, my grading was too tough. When I handed back exams, I caused some disappointment and anger.

Halfway through the class, I ran into a senior who had taken a graduate level class with me during the previous semester. He was a friend of five of the students in the undergraduate class, and they had enrolled on his recommendation.

"My friends are angry with me," he said. "Couldn't you lighten up for my sake?" I made a joke and told him that lightening up wasn't in my future plan. In restrospect, I should have taken his advice seriously.

The capstone experience for the class was a group project on an environmental topic. The class broke into groups of three and four students. Although I had written a list of suggested topics, I encouraged students to come up with their own topic. But out of the fifteen groups in the class not one of them did so. I was disappointed by this lack of academic independence. In retrospect, I think that they were simply trying to play it safe. There was always a certain risk that I would think poorly of a topic that wasn't already on my list.

Each group was also allotted fifteen minutes of class time to present their material, with no restrictions on presentation format. There was an incentive beyond the grade (at least I thought it was an incentive) to the project and presentation: The group with the best overall project and presentation was invited to my home for dinner.

Early on, a few of the students complained about the group nature of the project. There were worries about potential personality conflicts and the possibility of one person screwing it up for everyone else. But I thought that they would learn more by working together. To make sure that the groups were working effectively, I asked the students to write progress reports along the way. The reports I received suggested that the students' fears about the project were not materializing.

There was one strange twist, however. The night before the project was due, at about 10:30 P.M., someone knocked on my door at home. I was already in bed reading, so I put on my bathrobe to answer it. I thought it might be a neighbor. Instead, it was a student who wanted to talk about the length of his section of his group report. Each student was supposed to contribute a five to seven-page section in addition to a five to seven page group synthesis. The student standing in front of me had a question about these guidelines.

"Is it OK if my section is longer than five to seven pages?" he asked.

"How much longer is it, Craig?" I asked. Craig was a senior, short and muscular with a lot of nervous energy. A few weeks before, I had run into him on the quad and we had discussed what he was going to do after he graduated. He had told me that he had lined up a job on the currency futures market in Chicago. I thought that Craig's intense nature would serve him well in a fast-paced job.

Standing in front of me that night, Craig seemed to be in somewhat of a panic that he had gone overboard on his assignment and that I might dock him for not following the rules. I thought that this was kind of odd, but I also thought that I probably couldn't understand the mind of a twenty-one year old college student. I, of course, had been twenty-one years old at one time, but being panicky about assignments never was in my repertoire. As an undergraduate, I didn't care about grades. I wasn't concerned about getting ahead. Any sense that I had about achievement and recognition of achievement didn't come until my mid-twenties.

"About twelve pages, professor."

"Twelve pages? That sounds fine. Just hand it in tomorrow." Craig thanked me and left on his moped.

The next week I began to read the group projects. When I got to Craig's group, I read with more curiosity than usual. Craig's group had put together one of the better projects and it was generally well thought out and very impressive. All except for the portion written by Craig. His twelve-page portion was well written and well researched, but it had hardly anything to do with the other sections or the main topic of the project. There were a few sentences that seemed to be related to the project, but they were out of context in relation to the rest of the twelve pages. Even more strange, these sentences were highlighted in bold print in an unsuccessful and meager attempt to relate his portion to the rest of the project. Why write so much about something that was almost completely unrelated to the assignment? I thought about this for a little while and then it came to me. He had probably taken a report (hopefully his own) written for another class and slightly doctored it for his group project.

Now it all made a little sense. He had come to my home perhaps out of some fear that he was cheating and would get caught. Alternatively he had come in order to make me think that he had worked extensively on this project. If he wrote this paper for another course, then I wanted to find out what course it was. I went to the registrar's office and looked at his transcript.

He had taken a botany course two semesters previous and had received an A. The paper he gave me would have fit perfectly in a botany class. I called up the professor of that class.

"Did Craig X write a paper on the topic of Y?" I asked.

"Oh, yes," the professor said. "I remember that he wrote a very fine paper on that topic."

"Yes, I know it's a fine paper," I said. "I have a copy of it submitted for an assignment in my own class."

I thought about the situation. He had apparently written the paper himself. Of course, this paper was written for another course and he was trying to get credit twice for one piece of work. But technically, it was unclear to me that if this paper really was appropriate for both classes it could not be used for both classes. The problem was, however, that it was not appropriate for my class. It wasn't completely unrelated, but it was far from what I had asked for.

I decided to give Craig a grade of D. On his section I wrote, "well written and researched but not appropriate for the assignment." Craig never came to me afterwards to complain about his grade on the project. He just took the bad grade and moved on with the rest of his life.

I should note that I have had to deal with very few cheating incidents. It is true that I allow most of my classes to bring a crib sheet to exams. In these classes about the only way you can cheat is to copy from a neighbor. This has happened very rarely. Overall, my experiences suggest that college students are a pretty honest bunch.

My class continued on its path toward the final exam. The final phase of the class was the oral presentation portion of the projects. These were a mixed bag, but there were a few excellent pieces of work that seemed to make the experience worthwhile. I awarded my home-cooked meal at my house to one of the groups. We had a pleasant time together and I felt that I had ended my class on a good note.

I graded the final exam and handed out final grades. I decided to be generous in grading the final. By including the generally high homework scores, the average grade point in the class moved up to 3.1, but was still below my university's average for a class. Ten percent of the class received As. A key problem in my grading was that I wasn't aware that the "gentleman's C" of a decade ago had been inflated to a "gentleman's B." I received several calls and visits from people complaining about their grade.

I didn't mind the calls and visits. The students weren't too huffy and they generally took my comments in stride. The only unusual incident was when a student came into my office in a tank top with no bra, and said as she leaned over my desk, "I would do anything to get a better grade." It was easy to see what the "anything" was and I felt like I was caught in a B grade movie. I didn't know this at the time, but there is joke that begins almost with the exact same scenario. The professor responds to the come-on by asking, "Anything?" The student leans closer and whispers in his ear, "Yes, anything." The teacher then whispers back, "Well then, why don't you try studying?"

Unlike the professor in the joke, I felt too awkward to come up with a witty response. It's not that I was worried about a future sexual harassment suit. I just can't stand corruption, an aversion that I developed in my teens and twenties while working in the inherently corrupt business of construction. At any rate, the student took the news well that I wouldn't change her grade. This was the only time anyone tried to bribe me, either with sex or money, in my first seven years of teaching. So much for rampant sleazy intrigue between students and professors.

When I got back the student reviews for my class, they were more positive than I expected. Twenty percent did hate the class, but their comments generally were above the belt. The group of people who I thought had mixed emotions was not as large as I thought. I think that I was misinterpreting some students' reticence to participate during lecture as a sign that they had misgivings about the class. Many of my friends and colleagues have noted the relative passivity of students in the 1980s and 1990s. Fortunately, I've found that there are almost always a few students who are outspoken and liven up class atmosphere.

In the reviews, there were complaints about the boring and technical nature of science, the grading, and the workload. The number of complaints that related to a dislike of science broke down into two camps. There were those who did not like the problem-solving aspects of the class. The homework involved use of high school math, which turned off a lot of students. Problem solving involved examining details, and they weren't interested in the details. As noted earlier, there are many other classes in the sciences that do not focus on problem solving, and these students clearly would have preferred such a class.

The other camp would have hated any science class, details or no details. I know that, in general, American students think science is boring. Many books have been written on this topic and I have read some of them. The bottom line is that students feel negative about science in high school, and by the time they go to college, this view of science has hardened. So if someone thinks that science is boring before they get to college, I'm going to let myself off the hook and say that there is little I can do about this. I cannot, in one class, erase the effects of negative feelings toward science obtained from society and in high school biology, chemistry, or physics.

In terms of the complaint that the course demanded too much work, the student evaluations were fairly interesting. The students that complained about the workload reported that they spent an average of five hours per week studying outside of class. This meant that there was a significant disparity between what those students and I thought was a reasonable level of commitment to class study. It also probably meant that many university classes required scant outside work.

Many of the review comments were positive and some were enthusiastic, but overall the students gave the class average marks. Whatever the level of the ratings, students did not recommend this class to their friends. The next time I taught the class there were 30 percent less students. In fact, over the next six years each successive class brought significantly lower enrollment. I tried to counter the slide in enrollment by tinkering extensively with the class. I changed its title, changed homework, removed the group project and changed texts. But nothing worked. I was losing badly to the competition of other introductory science classes.

In the sixth and final year I added a required laboratory section. In doing so, I thought that the class would benefit because we could run experiments and do some fieldwork. The change may have improved class content, but it made the class even less attractive because having a lab meant that students would have to spend an extra hour in the classroom. Enrollment plummeted to nine students. I had big ideas, but their implementation was not practical. In making most of my changes, I was foolishly trying to increase enrollment by improving the class content. I was avoiding the obvious. If I wanted reasonable enrollment in my undergraduate classes, I had to lower my standards.

Finally, after six years, I decided that I had to completely revamp my teaching approach for undergraduates. I didn't want to teach to nearly empty classrooms and neither did Duke. At about this time, my university came to the realization that the era of exponential growth in higher education had ended. In response, the administration began to examine its balance sheets more carefully. They began a serious accounting of anything that could be quantified. One facet of a university that can be quantified easily is student enrollment. In its accounting of student enrollment, our administration found my department lacking. We were teaching half as many students per faculty member as a typical department. We were under the gun to increase our numbers.

So I made a bold retreat. I would now recognize the changes that had taken place over the years and instead of teaching students how to perform science, I would teach them how to appreciate science. I would now largely ignore detail, focus almost entirely on the big picture, reduce the workload, and grade easier. In upper-division courses, I would give up the notion of training students for jobs or research in science. In introductory science courses, I would drop homework assignments and material that assumed knowledge of high school science and math. My major goal had become less ambitious and more pragmatic. I wanted my students to be well-informed citizens who could think somewhat critically about scientific issues that affected society.

I employed this approach for the first time in an upper-division science class in which almost all students were science majors. All through the semester, I kept it easy and relatively light. To keep myself from feeling too bad about selling out my educational principles, I kept reciting the phrase "informed citizen" to myself like a mantra. I hoped that none of these students would go on to graduate school in the sciences. I felt that I had abrogated my responsibility to train them to be scientists.

The reaction to this new approach was enthusiastic. I gave lots of

As (to about 30 percent of the class), which I'm sure helped matters. I reduced the workload by about half and reduced the amount of material by about a third relative to previous classes. I did not completely cave in. I still required them to write short essay exams. I still required them to think about what they had heard in class and read in the text, but a lot less thoughtfully than before. Basically, I decided to stretch them only a tiny bit. If my classes were a track and field high jump, what I did was lower the bar a good meter. To do well, they did not have to work as hard or be as talented. Instead of having to work the "onerous" five to ten hours per week I previously expected in an upper-division class, they worked an average of three hours a week.

In reading the student evaluations of this class, it was clear that I was now much more in line with other classes at my university. The students noted that the class made them think and that it wasn't the usual memorize and spit it out science class. These students were science majors so I didn't have to get over the barrier of convincing them that science was interesting.

They also noted that they enjoyed the dynamics of the class. This seemed odd to me. I didn't ask for or receive much participation from this particular class. There were jokes that went back and forth between myself and the students about class logistics (tests, field trips, etc.) and issues of the day that had nothing to do with the class material. But little substantive interaction was going on. I think what they were saying was that they appreciated the lightheartedness with which I approached the class. I think that they also appreciated the dynamics associated with a "sit back and learn" approach to education.

After I read the reviews, I remembered a performance evaluation that I received the year previous from the committee that approved my tenure. They noted that I was considered demanding in the classroom, and characterized my teaching performance as "average." They, however, expressed "confidence that he will learn the skills necessary to become an excellent teacher." I laughed to myself thinking about this comment. The committee was right. I was a ridiculously slow learner, but I had learned the value of changing with the times and teaching easier.

So I'll play devil's advocate here and say what is wrong with having a happy classroom full of students? Sure, I only give them half to two-thirds of what I would like to give them. Sure, the material I've dropped is the most difficult. But the students learn something, don't they? And they like learning it, don't they? Won't they retain more of what they have learned this way? Isn't this better than having a classroom with too few students?

My answer to the devil is that yes, given the generally low expectations we have for our students, this is the better way to teach. But if we raised our standards, the students could do the work. With high standards present across the university, they would not think twice about putting in five hours a week outside of class or ten hours a week for that matter. They might be able to enjoy a well-taught class that required them to work hard because that class would represent the expectation of the university at large. This would not be a novel approach. We expected more of our students in the recent past.

And what about the trend of teaching science appreciation instead of science methods? Over the years, I've warmed to this approach for nonscience majors. These students tend to come to college with such negative feelings about science that we are pretending if we think that they will retain anything of value from a traditional science class. Classes that focus on scientific literacy, however, need to establish high academic standards. Also, this approach has spread to the teaching of science majors, and we are now producing, in small but worrisome numbers, a breed of science student that can't solve science problems.

At a place like my university, we get bright students and probably more than a third of them are also motivated and mature when they arrive. Another third, in the context of a university with high standards, would get up to speed and develop their motivation and maturity. But we are generally not challenging students at most universities, including my own. We let them coast if they choose to do so.

I feel bad that I cannot challenge the best. Imagine having an Olympics where the standards for gold medals are easily attained. For example, imagine a high jump competition where we grant a gold medal to every participant who can jump over the bar when it is at a height of one meter. The best get a gold medal along with many others, but they never find out just how much they can achieve.

At my university, a significant number of students are self-motivated and set their own high standards. Many of them set standards much higher than my own. But intellectual achievement should not be a self-service operation, particularly given the cost of higher education today. The American university could easily be a place where all students are expected to enhance their intellectual talents.

3

The Prestige Business

I was on a plane from Dallas to Denver to attend a conference of geologists. I was in an aisle seat, and on the other side of the aisle sat three women about a half a dozen years older than I. They were obviously traveling together. I could tell that one of them—the one in the aisle seat—had done a lot of traveling, but for the other two this trip was something of an adventure. I'm one of those people who likes to talk to total strangers on planes. It's a pleasant environment in which to find out about the variety of people's jobs and their lives. So I struck up a conversation with the women across the aisle.

The women were returning home from a jewelry show in Dallas and lived in a small town near Fort Collins, Colorado. The one sitting on the aisle was the owner of a small firm that made western style earrings and necklaces. The jewelry show had gone well and she had received many orders. I looked at the jewelry she was wearing. I am not a connoisseur of western clothing or jewelry, but her necklace of cast silver and carnelian looked distinctive. I hazarded a guess that she knew her clientele and had a good deal of talent.

One of the women asked me what I did. "I'm a university professor in hydrology," I said. Most people have never heard of the field of hydrology and these women were no exception. I explained what hydrology was.

"So you're a science professor," the jewelry designer said. She smiled. "You know, we could all use a little bit of help from someone like you." It turned out that all three of them had children in high school who had to design and create science projects. They wanted some ideas for science projects. I threw out a few ideas, but I could tell that they weren't too well received.

"Where do you teach?" the one in the center asked. I told them.

"That's a good school," the jewelry designer said. "That's one of the schools my daughter would like to go to." I've heard statements like this from other parents. In my conversations, they often say that their principal concern is that college degrees provide their children with opportunities for success. The perception is that schools with high reputations provide graduates with distinct advantages for admission to professional schools (such as medical school) and access to prestigious jobs. I agree with this perception, but I don't like what it implies about the world in which we live. If it is correct, it means that what a student learns in college is probably less important than where a student learns.

"Is she about to graduate?" I asked.

"She's a junior. I tell her that she is going to have to work hard and keep up her grades if she wants to get into a good school."

"Well maybe I'll see her in a class of mine in two years," I said.

"I hope that you do," she said. I wished her the best of luck with her daughter's college applications.

"There are a lot of good schools out there," I said. "I know that she'll get a good education."

It was one year later and I was on another plane, this time from San Francisco to Washington, D.C. Like most coast-to-coast flights in the summer, it was packed. The passengers had a slightly urban and sophisticated air. Generally, these are not the kind of flights where I can talk to my neighbors. The passengers read books, go through business reports, or tap away on their laptops. A woman in her sixties, well dressed and coifed, sat next to me, book in hand. I started up a conversation and she put her book away. She was a former high school principal who for the last fifteen years had worked as a private consultant to high school students in a major urban area. She helped them select the colleges that suited their interests, personalities, and academic skills. She helped write their applications. When she wasn't consulting, she visited schools and talked to college deans and admissions officers to make herself more knowledgeable. I'd never heard of a job such as hers.

"Are there many people who do this?" I asked.

"When I first began, there were only a few. Now there are quite a lot in the area."

"They work full time with high school students helping them with their college decision?" I sounded amazed that anyone could make a living doing this. I thought about the jewelry designer on the flight to Denver. Her daughter would not likely use the services of a college consultant. I thought of myself twenty years before. The cost differential between public and private education wasn't very great (about $2500 per year) and my family could have afforded the extra money. I hitchhiked to Massachusetts and checked out a couple of private schools the summer before my sophomore year in high school. In the end, I applied only to the local commuter school and the land grant state university seventy miles away. I didn't worry that certain career doors would not be open to me because I chose a state school. I wasn't even aware that such doors existed.

"Yes, it's a full-time occupation," she said, more than a little irritated with me. But I wanted to know more so I pressed on. Typically, high school students came to her when they were freshmen or sophomores. She interviewed them extensively to explore their interests and personalities. She told them what colleges look for in a high school transcript. She helped them select training classes for the SAT. They visited her a few times a year to discuss progress and choose appropriate colleges. "I've placed quite a few of my students into their college of first choice," she said, and I had no reason to doubt her. There was a market of high school students with wealthy parents who wished to optimize the chances of getting into particular schools. I am sure that this woman, just like the jewelry designer on the flight to Denver, knew her clientele and had a good deal of talent.

The difference between college selection on the part of parents and students twenty-five years ago and that of today is tremendous. More than ever, many students feel pressure to get into a school that is perceived to be prestigious. While "college consultants" are probably only used by a very small percentage of college applicants, the attention paid by students and parents toward the admission process has increased to levels that I honestly have a hard time comprehending. An entire industry has grown around the process of college selection and admission. Students routinely take SAT preparation classes. These classes were virtually nonexistent twenty-five years ago. A college guide well received by the public can be assured of large perennial sales. *U.S. News and World Report* derives a sizable revenue from its annual rating of colleges. Students in my hometown's local high school now apply to an average of six colleges. They and their parents are increasingly savvy and sophisticated in their selection process.

Perhaps the increased sophistication is the result of parents' experience. Mine was the last generation of college students whose parents did not commonly have a college degree. When I applied to school, my parents, neither of whom had been to a university, were mostly concerned that I go to a college of any stripe. Sure, they had their preferences. My father wanted me to go to West Point, thinking that some military discipline would do me good (who knows, he might have been right). My mother expressed an interest in a private school because she felt that somehow private schools were better. But neither of them were knowledgeable about universities nor could they give me any practical advice about university life.

Today's parents of students applying to college are different. They generally know from experience what college is about. They also know that since so many people today have college degrees, increased attention is paid to the reputation of the college where the degree is earned. Finally, they and their children know that good jobs for new graduates are not as plentiful as they were in previous generations (although opportunities increased dramatically in 1997 and 1998, and perhaps this trend will continue). They are concerned about obtaining any competitive edge that they can through their choice of college. For those twenty or so private institutions and five or so public institutions lucky enough to be viewed as prestigious (by those who pay for or attend college), the benefits of this increased attention paid to college selection are considerable. These institutions have the luxury of choosing from a large applicant pool. The private institutions also manage to find an applicant pool that does not balk at the prospect of paying over $100,000 for a college degree.

The increased attention paid to the relative prestige of institutions has in turn placed tremendous pressure upon universities. Those already deemed prestigious feel pressure to maintain their status. Those not in the upper tier have an overriding goal to attain this status. To be honest, not all of this pressure is external. A good part of it is driven by our own desire to feel important and compete with other universities for competition's sake. Whatever is gained through these efforts, however, is counterbalanced by the incredible amount of money and energy necessary to obtain and maintain the components of prestige.

There are at least two principal components to having a successful elite enterprise. One is to develop an aura of exclusivity and desirability. This effort is largely one of marketing and use of psychology. Through the years in response to the increased consumer savvy of parents and students, private universities have become increasingly sophisticated at marketing their product. Every year each university sends out its brochures to prospective students (750,000 brochures per year from Duke) and detailed information books to high school guidance counselors to broadcast its prestige and unique personality. Then there is the manufactured aura associated with the application process. Applicants are made to feel that they are applying to be part of a select and special group and that their academic records and application essays will be examined like jewels for both flaws and brilliance.

Ultimately, there must be some substance to back up the marketing. The facilities, the faculty, and the services a university provides must all be equal to the best in the nation if it is going to be perceived as one of the elite. The financial boom of the Golden Age allowed upper echelon institutions to upwardly redefine the elements of an elite university. They competed against each other to provide the best and be perceived as the best with little regard to cost. By the end of the Golden Age, they had built a massive enterprise that represented the combined desires and wishes of both the paying public and university faculty and administration. The non-elite private colleges and public universities partially followed step during this time period. In the post-Golden Age everyone is straining mightily to maintain their grandeur.

Let us look at some of the elements of today's typical private research university. First, the buildings. They are the primary symbol of the wealth and stability of a university. So there is a strong tendency to give universities the look and feel of an enterprise that has been in existence since time immemorial and shows every sign that it will enjoy continued prosperity. After all, when a student receives his or her degree they want some assurance that the degree will mean something in the years to come. We may be able to provide the same education a lot cheaper on a campus full of trailers, but no one wants a university to look like a trailer park.

Public and private universities went through an explosive period of building during the Golden Age. What they tried to ignore is that a building not only costs money to be constructed. It also costs money to maintain. The costs don't just involve paying for utilities, painting, and repairs. Eventually, they involve remodeling. For example, the main core of buildings on my campus was built in the 1920s and 1930s in a Gothic architectural style designed to give the university the look and feel of an old European university. These buildings were built to last at least a couple of centuries. The exterior walls are composed of a locally quarried stone and the exterior windows and doors are framed with limestone from Indiana. But after sixty or so years of use, even these impressive buildings are in need of remodeling.

My office is located in one of these buildings and was extensively remodeled from 1989 to 1991. At the time Duke was still ebullient about its finances. We spared no expense in remodeling the building and used top quality materials throughout. The effort put into modernizing my building, yet maintaining its stately feel, cost in excess of six million dollars. This is about the same cost as replacing the building altogether with a modern, but not stately looking structure. With the end of the Golden Age, our economic outlook has become much more cautious and most remodeling of other buildings at Duke has been tabled. Such tabling is called deferred maintenance. As of 1998, my university had accumulated about $30,000,000 of deferred maintenance. This is small potatoes, however, in comparison to some institutions. Yale University is probably the most notable on this front, and has had to confront more than one billion dollars in deferred maintenance.

Despite the backlog of deferred maintenance on campuses across the country, universities continue to build new structures, often in grand style. The motivation to create these new buildings comes from many sides. For example, wealthy alumni who wish to give money like the idea of having their name on new buildings. But most of our need to build comes from an idea widely held by university administrators and faculty that growth is essential to a university's health.

Next look at faculty. Parents, students, and university administrators want world-class scholars to inhabit prestigious institutions. World-class scholars, however, if they spend all of their time in the classroom, will not be world-class scholars for very long. They need resources and time to perform research. For those in the sciences, such resources (labs and lab equipment) typically cost hundreds of thousands of dollars per faculty member. Much of the money for research equipment has come from federal grants. But grants do not fully pay for all of the equipment or for all of the time needed for research.

In the Golden Age, universities, public and private, competed against each other to create more time for faculty members to perform research. Elite universities cut classroom time from nine to twelve hours per week per faculty member to three to six hours per week. Since student enrollment did not decrease during this time (in fact, it increased markedly), the number of faculty had to more than double nationwide to make up for the teaching shortfall. The demand for additional faculty fueled increases in faculty salaries. And to manage all of these additional faculty, universities hired more senior and junior administrators. Of course as seems to be the way of all administration, university administrative staffs across the nation grew at a more rapid rate than the professorate. Administrative staffs swelled to such a degree that new national organizations of university administrators formed during and following the Golden Age. Some of the names of these organizations are ripe for parody such as the National Association of Presidential Assistants in Higher Education. Presumably, the head of this organization is the "President of Presidential Assistants."

Universities, in an aggressive effort to elevate their intellectual stature, also routinely began to raid each other of prestigious scholars. These raids were often a zero sum game (all but a few institutions lost as many faculty as they gained) but they served to raise the salaries of star quality academicians more dramatically than that of other faculty. Furthermore, the newly arrived star faculty usually negotiated teaching loads lower than the rest of the faculty. Some were not required to teach at all.

In terms of faculty and administration, the Golden Age produced a long list of high-salaried employees. In the 1990s, the associated payroll has become burdensome. Because faculty and administrators carry a good deal of clout, the number of these positions has generally not been reduced. The end of the Golden Age has seen universities try to partially control salaries by replacing retiring faculty with part-time and finite-term contract (as opposed to tenured and tenurable) full-time instructors. Salary increases in this decade have been below that of inflation at many universities. It is easy to imagine that in the not so distant future, faculty salary increases nationwide will consistently lag behind inflation and perhaps at many universities, the number of faculty (and hopefully, the number of administrators) will be adjusted downward.

Now let's look at the staff. Until the 1970s, private universities and colleges were allowed, because of their nonprofit status, to hire staff below the federal minimum wage. They widely took advantage of this legal loophole. For example, in reference to the downright unconscionable wages of that time, my university was referred to as "The Plantation" by local black citizens (it is still sometimes referred to by that name). The low salaries allowed private universities nationwide to hire large numbers of maintenance staff. Since that time, wages and benefits have dramatically increased (although private universities and colleges still tend to offer relatively low salaries to staff) and so have the total costs. These employees have the least clout and it is not surprising that in response to post-Golden Age budgetary constraints, there has been pressure to reduce maintenance and secretarial staff across the nation.

Finally, let's look at the services oriented to students and education. Over the last twenty years universities and colleges have increased staff devoted to academic advising, tutoring, psychological counseling, and job placement. They have added teaching and learning centers to give professors and teaching assistants guidance on how to teach more effectively. They have large computer systems for students that need to be updated every three years or so to accommodate rapid changes in technology. They have added multimedia capabilities to classrooms that also need frequent updating. They have built new athletic and recreational facilities. They have remodeled dining halls. All of these incremental changes are considered desirable by students and parents of students. Universities and colleges have yet to curtail spending in these areas significantly. But it is likely that in the future they will do so.

After two decades of increasing tuition at rates well above that of inflation and several decades of dramatic growth in federal spending on education, private universities nationwide have built an enterprise that is in many ways quite admirable. They have an impressive look and feel. More important, they have produced some palpable benefits to society. They produce scholarship that in the sciences is partly responsible for innovation that fuels our country and in the humanities gives us some perspective on the world in which we live. They provide students with access to much of the latest in technology and guidance to help prepare them for their future and make them aware of all that the university has to offer. Ignoring their increasingly modest expectations of students for the moment, it is worth noting that what they teach is often of high quality. The pursuit of growth with little regard for cost has been mimicked to varying degrees by public institutions.

The price tag associated with these achievements at both private and public universities is intimidating. Maintenance of the entire enterprise requires a heavy dependence on growth in federal funding and tuition. Growth in federal funding of higher education in terms of inflation adjusted dollars has come to a halt at the end of the 1980s. If growth resumes (and there are indications that we are entering a mini-boom in federal spending, particularly in health science related university research) it is highly improbable that it will be sustained for decades as it was during the Golden Age. With regard to tuition, private universities have already priced themselves out of the range of affordability for most Americans. The cost of four years of education at an elite private university is now greater than the cost of an average house.

Ironically, the rise in tuition is partly the result of financial aid. Almost all of the elite universities practice "need blind admissions." They admit students without consideration of their ability to pay and ultimately the financial aid costs associated with such a policy are imbedded in the amount of tuition charged. If a student comes from a family of modest economic means and has an excellent academic record, universities promise to provide the financial means necessary to ensure that he or she can enroll.

Roughly 20 to 30 percent of tuition at private universities that practice need blind admissions goes toward financial aid. As tuition rises due to increased university costs, so do the number of people needing financial aid which in turn causes tuition to rise even higher. This upward spiral of tuition and financial aid has caused almost all of the less wealthy private universities and colleges to drop the practice of need blind admissions altogether. It should be noted that for most Americans need blind admissions assures affordability in only an abstract sense. Students and parents of students may be eligible for loans of tens of thousands of dollars, but they often find loans of this magnitude too daunting. As a result, the demographics of the applicant pools of the elite universities, traditionally biased to the privileged class, have shifted even more toward the wealthy over the past two decades.

In the past, university senior administrators across the country had been unwilling to admit that the costs associated with growth and elevating the components of prestige have been too great. For example, a few years into my career at Duke, one senior administrator tried to justify tuition costs by telling me that over the last twenty years tuition had roughly tracked the cost of a typical middle-class car. Not only is this not true, but the car owner of today holds onto a car considerably longer than the car owner of twenty years ago. More recently, some college presidents have begun to publicly acknowledge that we have created institutions that are financially burdened and increasingly unaffordable to the middle class. The talk has not translated into much action. Attempts at curtailing costs and tuition hikes have been very modest. Dramatic increases in the stock market and in private donations in the 1990s have given many universities a good deal of economic breathing room. Universities continue to try to build and grow and blithely hope that either through continued dramatic increases in endowment or growth in federal funding the Golden Age will return.

Their lack of interest in cutting costs is at least partly borne of an inflated view of their own importance, but it is also partly motivated by the continued demand for prestige shown by the public. Sometimes, when I meet parents of Duke students, I talk to them about college costs. I find it surprising that they are usually without complaint. They like what we offer their children and fully accept that our wide array of services costs large amounts of money to maintain. Some of them oddly boast about the sums of money that they are paying or the size of their loans. I am aware that I am talking to the small subset of the American population that has chosen to pay big-time dollars for their children's education. But there continue to be many more students who have parents willing to pay than we can house. As long as there remains a large demand for entry into prestigious universities, the pursuit of prestige and its associated economic demands will likely continue.

4

Shortening the Yellow Brick Road

As a child, I spent one day of every other week at my grandfather's automotive junkyard. I had a great time wandering through the abandoned Chevys and Cadillacs. I especially liked the cars from the fifties with their hood ornaments, fins, and Jayne Mansfield bumpers. Most of the cars were heavily damaged and were delivered by tow truck. My grandfather collected misplaced keys for these cars and kept them in piles placed in hubcaps in his office.

We had an ongoing game with these keys. If I could find the trunk key of any car, I could keep whatever was inside. I spent a fair amount of time matching trunk keys with cars. It was surprising what was left in these trunks. There were whole sets of golf clubs and basketballs. There were shoes in good condition (I didn't care about any shoes). And once there was the entire collection of Frank Baum's *The Wizard of Oz* series in hardcover. I vaguely remember that it was my uncle who found these books.

I wasn't much of a reader at the time (my interest in reading did not perk up until after I left high school), but I enjoyed these books immensely. They had a plain, midwestern style that appealed to me. And then there was Dorothy. I was direct and forthright as a child. Dorothy, although a girl (a definite negative for me at the time), was direct and forthright. She had a dog. I wanted a dog. It was a match made in heaven.

Virtually everyone in America knows the first book in the Oz series either from reading it or from watching the movie version. Dorothy and her dog Toto follow the Yellow Brick Road with their three companions. They all want something from the Wizard, and they visit him in his Emerald City. He is an imposing mysterious figure, but Dorothy and Toto manage to quickly expose the Wizard as a pompous, ordinary man. Despite his shortcomings, he does know how, through almost transparent artifice, to satisfy their requests. The lion gets a medal as a sign of his courage, the tin man a watch for a heart, and the scarecrow a document (a testimonial) to certify his intelligence.

Since the 1980s, the American university has been taken to task (perhaps most conspicuously by President Reagan's Secretary of Education William Bennett) for its Oz-like approach to undergraduate education. The argument has been made that while we have imposing edifices and speak in lofty tones about the quality of our education, the diplomas we give represent no more education than the testimonial received by the scarecrow in the Emerald City.

This criticism is mostly directed at the absence of required courses in western civilization in our nation's college curriculum. I leave this ideological issue as one for politicians and those in the humanities to fight about. However, coincident with the elimination of required courses in western civilization over the last thirty years there have been significant reductions in general expectations for graduation. That we have made individual classes easier is the subject of an earlier chapter. We have also pruned classroom hours. By doing this, we have explicitly discounted the importance of education in college life. Receiving a university diploma still requires a significant amount of education, certainly far more than that found at Oz University. However, it is clear that students are spending less time studying and in class than recent generations.

For example, take my university. In 1968, classes took place six days of the week. Students were required to take 124 credits for graduation. For students in the humanities and social sciences (where laboratory sessions are uncommon and a typical class has three hours of lecture per week), 124 credits amounted to about fifteen to sixteen hours per week in class plus required courses in physical education. For those in the sciences, required laboratory sessions increased the class hour load an additional two hours. During the period 1968 to 1986 the number of class hours per week required for graduation was reduced to about thirteen for those in the humanities and social sciences and to about fifteen for those in the sciences. Since the 1980s it has been very common for students to have two to three courses worth of advanced placement credits from high school. So students only need to enroll in twelve hours of class per week in the social sciences and slightly more in the sciences. All in all, student class hours required for graduation have been reduced by about 20 to 25 percent.

Given that we have both reduced class hours and lowered our standards for student performance, students here today are working significantly less than the students of the 1960s. It should be noted that the semiquantitative assessment of the reduction in workload given above agrees with the evolution of student views of the university. In the 1960s, students referred to Duke, noted for its striking Gothic architecture and heavy workload, as the "Gothic Rock Pile." Nowadays, students affectionately refer to the university as the "Gothic Wonderland."

Why do we require less? I cannot come close to fully answering this question. There are many contributing reasons. Probably the most influential is the political shift to the left of professors and administrators after the Vietnam War. Professors, deans, and presidents began to question traditional standards of achievement and allowed students a greater say in the running of the university. If students complained that they were being made to work too hard, then faculty and administrators felt compelled to reduce the workload.

Another contributing reason has its source in society's overall expectations. We live in a much more permissive society and we have tended to reduce the requirements for achieving life goals. Universities are simply following cultural fashion by reducing requirements.

A reduction of workload also has a palpable benefit for professors. Making students work less means more time for a professor to perform research. A student body that spends one fifth less time in the classroom also creates a professorate that spends one fifth less time in the classroom. It is a change that has proved "beneficial" to both student and professor because students are generally happy to spend more time at play.

It is also important to note that universities' reduction in workload, both course hours and class material, has only rarely been met by significant cries of protest from those outside of the university. The political right and left are much more concerned with the ideological content of our courses. Professional schools continue to accept our graduates in large numbers. Corporations still come to hire our new graduates. Those hiring students in the science and engineering fields are not complaining about their lack of knowledge. Those who hire students in other sectors, the majority of hiring, seem to be concerned only that our students are personable, receive high grades, and have a diploma. So another reason we've reduced workloads is that neither parents, politicians nor corporations seem to care whether we do or don't. As long as we provide students with a diploma and as long as we don't spend too much time indoctrinating students with radical political philosophy, they seem to be quite happy with the education (if not its cost) that we provide.

Requirements for departmental majors have also been made easier, especially for departments that have low enrollment histories. This too has created more free time for students. Departments have made these changes because of increased pressure to increase their enrollments. Faced with budgetary constraints at the end of the Golden Age, university administrators across the country began viewing departments as if they were divisions in a corporation. Academic departments were made fully aware of their balance sheets where revenues came from the number of students taught and the amount of grant money received. I became acquainted with the effect of the new economics on departmental academic standards as a member of the Curriculum Committee of my school.

Our charge on this committee of seven or so professors in humanities, social sciences, and natural sciences was to review requests by departments to create new major areas of study or change requirements for majors. We made recommendations and passed them on to the Faculty Council, which gave final approval.

An example of the erosion of requirements for major areas of study was a request by the Department of Classics to split their existing major into two. Normally we summarily rejected such requests. The general feeling on the committee was that students already had many choices for majors. Splitting a major into two usually meant a narrowing of scope of each major, and only rarely was this considered beneficial.

The department was requesting two majors in classics, classical languages and classical studies. The proposed classical languages major was to be the same as the existing classics major. Through class work, students were required to study Latin and Greek and to read major works in these languages. The classical studies major was to be simply a watered-down version of the existing major. Students were not required to learn any Latin or Greek. Instead they would read the major works in translation. The justification made for this change was that many excellent translations of all major works are available and, hence, it is not necessary for a student to read them in their original language.

Although I knew no Latin or Greek, I had a personal soft spot for classics as an area of study. Students in classics were following a path that formed the historical foundation of college education in the United States. Until this century, college education consisted largely of courses in Greek and Latin and on the major philosophers or historians of the Greeks and Romans. Until a decade or so after the Civil War, these courses and mathematics were required of everyone at most colleges. The great and not so great political decision-makers of our nation's past had been educated in the classics. This major was perhaps our only significant link to the education of previous generations.

We discussed this proposed change. One of our committee members was somewhat indignant about the proposed major that lacked a language requirement. "If one seriously studies German culture, French culture, Spanish culture, or any culture, one needs to learn the language of that culture," he said. "Translations cannot possibly convey the full spirit of a text written in its original language." We nodded our heads in agreement. There were no letters from university administrators expressing their support for the new major. On its own, without university administration support, the change was dead on arrival.

Then another professor asked, "The justification for this new major is so weak. There must be something else going on. Does anyone know the real reason classics is proposing to do this?"

The chair of the committee took his elbows off of the conference table and sat fully upright. He was tall, athletic and had an easy smile. "I talked to the chair of the department about this proposal," he said. "They are having a serious problem with enrollments. They are hoping that this change will increase their majors."

"How many majors do they have?" I asked.

"Less than twenty, I believe," said the chair of the committee.

Now everyone understood this proposal. This venerable major, the former foundation of undergraduate education, had fallen on hard times. The classics department was desperate enough to come up with the idea that one could major in classics without any knowledge of a classical language.

We discussed the change further. In light of the information that classics had very few majors, the tone of the discussion was far more sympathetic. We were no longer a committee trying to vouchsafe the integrity of the university's curriculum. Instead, we were trying to ensure the health of a department in the eyes of the university administration. At the end of the Golden Age, universities no longer felt a need to represent the entire breadth of scholarly pursuits. Perhaps, since classics departments had recently been eliminated in some schools, our university administration would feel compelled to reduce classics' faculty size or eliminate the department if enrollments stayed low.

"Do you really think this change will help enrollments all that much?" I asked. My feeling was that a painless, language-free approach to the classics was not going to prove to be significantly more attractive than the existing major. In an open marketplace of majors, classics just wasn't attractive enough to the student of today.

"Well, we obviously don't know," a professor in the social sciences said. "But if the classics department wishes to try this approach, we ought to give them a chance." This sentiment carried the day. We voted unanimously to approve the classics department's proposal to split their major into two. Collectively, we hoped that they would succeed in significantly increasing their enrollments through such a change.

I don't regret voting as I did. I do regret that departments feel compelled to weaken requirements in an effort to appear more attractive. There are only a fixed number of students. Increases in enrollments in one department come at the expense of decreases elsewhere. If a department is successful in increasing enrollments by reducing requirements, they provide an incentive for other departments to do the same. Over time departmental majors may begin to have the look and feel of television shows. They may be designed for broad appeal rather than for any intellectual purpose. The same can be said for the watering down of individual classes. I note that my own department also weakened its degree requirements considerably in the 1990s in an effort to increase enrollments.

The effects of making classes easier, reducing time in the classroom, and allowing students, if they wish, to avoid hard classes have combined to reduce college class work to the equivalent of a part-time job. A typical student will spend twelve to fourteen hours per week in class—if they attend class—and ten to fifteen hours per week studying. Granted, being a student is an inherently inefficient process in comparison to a real job. One has to shuttle around to different classes. One has to work on four or five different topics at once. But there are also built-in efficiencies to student life, especially at universities where students typically live on campus. The commute time to the classroom is small. Someone is cooking for you and doing much of your housework.

Some departments in the sciences and engineering still have fairly heavy class requirements for their majors, albeit somewhat less than those in the 1960s. Students in these departments still have many labs to attend in addition to their lectures. At my university they do not typically receive extra course credit for the laboratory portion of classes. The workload for individual classes that are expected for medical school or expected by employers in the science and technology sector is still heavy. Also, with the rise of college costs, there is an increasing trend of students having to balance their school with full-time or significant part-time employment. As a result, we have created a bimodal population of students. Many are busting their butts, but others likely have more free time than any college attending generation of the twentieth century.

While such a change has created greater opportunities for faculty to pursue research, it has put a great deal of pressure on universities to find nonacademic activities for its students. At Duke, we have responded to this pressure in a number of ways. For example, much effort is expended toward keeping our basketball team nationally competitive. Our team generates significant revenue and at the same time gives many students something exciting to watch. As I write this chapter, a new multimillion dollar state-of-the-art athletic recreational facility is being built on my campus.

The vacuum has also been filled with an increasing amount of partying and general carousing. I know from personal experience that the current generation of college students did not invent getting loaded at night. I also know that it is unrealistic and unwarranted to rigorously enforce a ban of underage drinking on campus. But the post-Golden Age students have elevated getting bombed to new levels. When I came to Duke, this type of activity was at its peak. Fraternities were allowed to distribute free beer on campus several evenings a week. As a result most evenings were filled with a level of drunkenness that made it hard to maintain significant standards in the classroom.

For example, during my first few years as a professor, I found out something curious about homework assignments and exams. If the due date for an assignment or the date for an exam was a Monday or Friday, the quality of the work was not infrequently abysmal. If the due date was on a Tuesday, Wednesday, or Thursday, student performance improved considerably. Students were spending so much time carousing Thursday through Sunday that except during the midweek it was not possible for some of them to do more than show up for class and take notes.

I was largely oblivious to the degree to which drinking and partying dominated campus nightlife when I first arrived. During my first year,

I was even foolish enough to have a homework deadline on the day following the final game of the NCAA basketball tournament (Duke's basketball team had made it to the final game). At the time, I thought that students would plan in advance and finish their homework ahead of time. In retrospect, this expectation was ridiculous for any university. I went to campus the night of the game. The alcohol was flowing freely, and jubilant students frolicked in front of a giant bonfire celebrating the first ever basketball championship for their beloved Blue Devils. It was fun to watch the revelry. After I watched the bonfire for a while, I walked along the road toward my office. A car stopped alongside me. The driver rolled down the window, his face red with the joy of victory and alcohol. I thought that he was simply going to scream something like "Blue Devils rule!" But instead he shouted, "Is the homework still due tomorrow?"

"No," I shouted back. "It can wait."

"Thanks," he said, and continued to weave down the road. How he had the presence of mind to recognize me through the alcoholic fog and remember about the homework due the next day I will never know.

Recently, my university began to recognize the problems created by the large amount of free time associated with the reduction of our requirements. This change was at least partially instigated by a speech given during Founders Day, an annual university-wide celebration. Founders Day is designed to give my university a sense of its glorious, if brief, history. Commencement would rank higher on the list of days to avoid controversy, but this day comes in at a solid second. One does not expect anything too meaty in the Founders Day speech. So it came as a huge surprise when in December 1992, one of our most nationally visible colleagues, Reynolds Price, used the opportunity of being invited to give the speech to clear the air about the state of our undergraduate education.

Mr. Price, a solidly built man with gray hair and a moonlike face, projects an air of confidence. He carries the credentials of a national literary figure, a long-time faculty member and an alumnus. It was expected that he would give the typical speech about the greatness of the university or perhaps about the intrinsic worth of higher education. Instead, he served a stiff cup of coffee.

His speech could have been dismissed as the musings of a curmudgeon with a sentimental view of the past. Price had contrasted his level of education as an undergraduate with those of students today and found today's students wanting. Back when he was a student at

Duke in the 1950s, the students, who at that time generally came from the South, were serious and worked hard. According to Price, these students represented the intellectual future of the South. Duke was a mecca for the southern student, a place where one could engage in topflight education and still stay in the heart of southern culture. Not only did students work hard because the professors had high standards, they worked hard because they wanted to learn. Discussions on great issues and ideas were common outside of the classroom.

In contrast, the culture of my university had stripped itself of its southern roots. Students came from all over the country, and we were no longer a special place in the South for southerners to send their best and brightest. With the loss of roots came a loss of intellectualism in the student body. Price had listened to students talk to each other before and after class and found that conversations commonly began with "You wouldn't believe how drunk I got last night." Students did not work hard and it was no longer expected that they do so.

Price noted that if he were to give the grades students really deserved, they wouldn't enroll in his classes regardless of the worth of the class or the notoriety of their professor. Since other professors gave strictly As for average work and Bs for below average work, he was all but forced to follow along. In the open marketplace of classes, he had to accede to low standards and give high grades or face an empty classroom. The picture he painted was bleak. My university, according to Price, had degraded to a party school devoid of intellectual purpose.

Usually university administrators get very defensive about such complaining because one of their primary concerns is to preserve the image of the university. For example, when former Secretary of Education William Bennett criticized American universities for their lack of intellectual standards, university presidents across the country soundly attacked him for being flippant and inaccurate.

So it was a surprise that in response to Reynolds Price's speech lambasting the decline of intellectual life, my university cautiously acknowledged that there was substance to what was said. Changes were made in campus life. The most profound of these was the curtailment of beer drinking. Partly in response to the Price speech (but probably also in response to fear of litigation from parents and complaints from the federal government about underage drinking and the general drunken state of campuses throughout the nation) drinking parties were limited to weekends. Beer kegs were restricted and dispensing of beer could only be done by university approved bartenders.

This change had a major impact on student attitudes about the university. Many students complained bitterly that there was little to do on campus except go to classes and study. Their disenchantment grew and culminated with a raucous protest under the light of a bonfire. Other campuses that have tried to curtail alcohol use have fared worse. Alcohol restrictions led to full-fledged riots at Michigan State and Washington State in 1998. Predictably, attempts to create alternative forms of entertainment have been ineffective.

The diminution of party life on my campus produced modest benefits for me. I could now make a due date for an assignment on any day but Monday and have a high expectation that it would be done at least reasonably well by almost everyone in class. University faculty members continued to inflate grades, and students typically weren't required to work very hard. Although bored, they still had plenty of hours to socialize, and paid a little more attention to their classes. We may not have a significantly more intellectual atmosphere, but we have ended up with a more sober one.

Removing beer on weekdays, however, is only a response to the symptoms of a problem. The real heart of the matter is not how much students drink, but our expectations of students' academic performance. There have been other times when educational standards in universities waned. Henry Adams wrote of his Harvard education in the 1850s:

> Any other education would have required a serious effort. But no one took Harvard College seriously. All went there because their friends went there, and the College was their ideal of social self-respect. . . . It taught little, and that little ill, but it left the mind open, free from bias, ignorant of facts, but docile.
> *The Education of Henry Adams*, 1918

To what degree this description of Harvard at the time is realistic or represents Henry Adams' natural predilection toward dyspepsia (a trait that I share) is unknown. There is likely some truth to these statements because Harvard underwent a massive reform of its educational standards after the Civil War. The current state of our educational standards is not nearly so poor as that described by Henry Adams in his time. Nor are our standards quite as bleak as those described by Reynolds Price. However, there is ample room for reforming our education and making our students work and think harder.

5

The Sports Machine

I'm a big fan of athletics, particularly baseball. I played a wide range of sports as a kid, but didn't have the talent to pursue anything beyond intramural sports in college. The sports section is the first part of the newspaper that I turn to every day during the baseball season. But I also pay attention to college basketball, especially now that I live in a part of the country where basketball dominates the sports section every winter. I will even watch the occasional college basketball game on TV if I can find the time. So in the winter of my first year, when I got a call from my dean asking if I wanted a pair of tickets to that night's basketball game, I gladly accepted.

I'm sitting in literally the worst seat in the house of my university's pride and joy, Cameron Indoor Stadium, watching the Duke men. I'm quite happy to be sitting where I am and so is my friend (another untenured professor, but from a different department). Cameron Indoor Stadium is old, has poor ventilation, and has uncomfortable seats. But it has plenty of character and with a capacity of only 9300, every seat in the house is close to the action, even mine. It is arguably the best place to watch college basketball in the country.

Unlike most universities with successful big-time basketball, Duke has not succumbed to the urge to build some large domed structure to honor the sports gods. (As I write this our neighbor university, North Carolina State, has just broken ground on a $120,000,000—plus likely cost overruns—21,000 seat basketball arena to replace a 12,000 seat basketball arena built in the 1950s.) Instead, my university has been steadfast in its devotion to its original basketball arena, a 1940s model. With so few seats available in an area of the country where college basketball is nearly a religion, tickets are as hard to find as hen's teeth, especially during a game with an Atlantic Coast Conference rival.

In the arena, I watch the fans as much as I watch the game. The ambience is purely electric. These fans aren't just watching the game in front of them. They are living the game. Many of them hang their emotions on every bounce of the ball, every pass, and every shot. It's like being at a major-league baseball game during a pennant drive in late September. The students, some of whom have camped out for a couple of days to get their tickets (students get in free, with tickets given out the evening of the game first come, first served), get the bleacher seats closest to the court. They stand throughout the game and are clearly the catalyst for the hyperbolic level of enthusiasm in the arena. In honor of their exuberance, and their well-orchestrated, sometimes obscene, and sometimes very funny and clever cheers, they have earned the name the "Cameron crazies." The coach of the basketball team calls these 3000 or so loud student fans his "sixth man." It's no wonder Duke rarely loses on its home court.

The remainder of the seats are sold predominately to members of the Duke University athletics booster club, affectionately named the Iron Dukes. At the time of my attendance at this game, supporters who joined after 1984 contributed a minimum of $2000 per year to the athletics department of the university in order to have the privilege to purchase a pair of season tickets at an additional cost of about $600. (In 1998, the minimum contribution required was raised to $3000 per year. Either amount buys a hell of a lot of iron.) Given that there are about fifteen to eighteen home games, the cost of a pair of tickets works out to about $170 including the contribution to the athletics department. The bigger the contribution, the better the seat and God only knows how much those sitting in the midcourt seats have paid. Let's just estimate that amount at $10,000 and that the average amount paid for the 2000 or so pairs of tickets available through the Iron Dukes is about $4000. So as a result of a successful basketball program, the athletics department is able to generate roughly eight million dollars in revenue from ticket sales and contributions alone. The few faculty attending the game that I am watching have either paid this contribution, received complementary tickets like me or are old enough to have begun purchasing season tickets when faculty did not have to contribute to the Iron Dukes and have been grandfathered in.

When one excludes research associated with the medical school at Duke, the amount of money raised through basketball attendance alone is equal to about 15 percent of the total amount of money received from research. Add in basketball merchandise sales, revenue from television contracts, and the NCAA basketball tournament in a successful year for the Atlantic Coast Conference and our basketball team is responsible for over ten million dollars in annual revenue. Then there are the intangibles associated with successful basketball like the happy alumnus who might be inclined to donate a little more money to the Duke University Endowment in response to a team appearance at the NCAA Tournament Final Four (a tournament held at the end of the season in which the top four teams in the country play to determine the national collegiate champion).

The large financial windfall associated with our basketball team is a phenomenon that began in earnest sometime in the 1970s. Universities like mine that have chosen to invest in scholarship athletics have also experienced significant revenue enhancement in basketball or football. The growth of college sports overlapped the time of the rise of research in the American university. Unlike federal support for university based research, however, interest in college athletics and the associated payoffs have not leveled off in the 1990s. If anything, interest levels and payoffs have skyrocketed during this decade. For example, the NCAA generates over $200,000,000 per year from television contracts associated with basketball.

While the dollars generated through college athletics are currently much less than those generated by university research nationwide, current trends suggest that they will outpace university research dollars at a few universities in the not so distant future. They may even outpace tuition revenue at a few state universities. Perhaps George Steinbrenner (the current principal owner of the New York Yankees baseball team) might make a good college president under these conditions.

Even with its current relatively small revenue (compared to university research and tuition), college athletics, particularly Division I NCAA football and basketball, have had a significant impact on the American university. On the plus side, it has provided students and alumni with a source of unity and entertainment. Given that we provide our students with so much free time, we might as well also provide them with something fun to do that does not put them at risk physically and encourages school spirit.

On the minus side, the financial pressures to create successful sports programs have also created a fair amount of corruption. In order to accommodate competitive football and basketball, universities across the nation must routinely lower their academic standards for athletes. That they have done so with the blessings of their alumni and (for state schools) their state legislators says a lot about how much we value athletics relative to academics.

My university's athletic program is considered to be a model of the viability of mixing athletics with quality education. I agree that we likely represent the best that can be done with scholarship athletics. Being among the best means that we expect our scholarship athletes to enroll mostly in real classes with real tests (more on this later), generally choose a major that represents a fairly legitimate area of study (in other words, we don't offer a major in something like leisure studies, a "subject area" that seems to be quite attractive among athletes at some schools with large scholarship athletic programs), and expect them to graduate (about 90 percent of our basketball and football players graduate within six years of entering Duke).

Although we are among the best in terms of reputation, we consistently offer scholarships to more than a few athletes that have no aptitude for academics. Whatever their prowess on the basketball court or football field, this type of athlete cannot possibly compete with our nonathlete students academically. A forty inch vertical leap seems to make up for a lack of academic ability. For example, the average SAT scores for students at my university (math and verbal combined) are in the range of 1350 to 1450. The average SAT scores for the basketball players hovered around 900 in the early to middle 1990s.

Our football team does significantly better but still falls far short of the general academic standards of the university. The average SAT score for the football players has been about 1080. This level of achievement on the SAT, while quite low for the university, is fairly respectable. It is comparable to the average score of the entire student bodies of many state schools such as the University of Massachusetts. It is also well above the scores for almost all other Division I scholarship football teams (Stanford and Northwestern are other institutions that have comparable SAT scores for their football teams).

Given that we don't stoop as low on the academic achievement ladder as other football teams in our conference, it is not too surprising that our football team, which was successful before the era of big money, does poorly today. Other schools opt for the talented and not so smart. We tend to opt for the smart and not so talented. Like those at Stanford and Northwestern, our football fans have to settle for a once a decade or so miracle winning season. We do stoop almost as low as others in basketball, however, and that, no doubt, is the major reason why we are almost always competitive with other basketball teams in our conference and in the nation. It also doesn't hurt to have one of the country's most outstanding basketball coaches.

Even before the age of big money in college athletics, big-time football and basketball generally represented the sleazy side of the American university. The amount of money involved was much smaller during the point-shaving basketball scandals of the fifties. But in every decade, NCAA football and basketball have always attracted the bribing recruiter, the athlete who didn't show up for classes, or the gambler who wanted a little edge. There is more than a bit of Las Vegas in college athletics, a mixture of slime and neon lights.

I first came into contact with the special aura of college athletics inadvertently, when I was seven years old. Virtually every year, my parents would drive us down to Miami Beach during school winter break. We would stay in rock-bottom cheap hotels on the beach (usually less than fifteen dollars a night). While my parents would spend their time talking to Jewish immigrants from cities like New York, Chicago, Cleveland, and Baltimore, my brother and I would be left to our own devices. We would wander around the pool, the beach, or maybe take the bus down to Lincoln Road and buy some trinkets with our pocket money. Imagine letting kids today take the bus on their own to south Miami Beach. College athletics and university life aren't the only things that have changed over the last thirty years.

Our visits coincided with the Orange Bowl football game and in those days college football teams also stayed in rock-bottom cheap hotels. When I was seven years old, my father's objective in hotel selection and the University of Nebraska football team's objective were in complete alignment. These huge midwestern farm boys invaded our hotel, and amidst the population of Jewish immigrants and their children, they appeared Goliath-like.

For a sports crazed child, the appearance of this football team was akin to manna from heaven. I quickly became a Nebraska football fan and cheerily hung around the players at the pool. Perhaps I hoped that exposure to large muscular people would somehow counteract the influence of my gene pool with its multiple generations of diminutive men, far more skilled at adding and subtracting than scoring points. On New Year's Eve, the night before the Orange Bowl, the football team was sequestered in its hotel wing. They were presumably getting rest for the game ahead. I wandered over to their wing of the hotel and they were partying away. The doors of their rooms were wide open and the alcohol was flowing.

I went into the room shared by one of the stars of the team. He sat in a folding chair, his eyes glazed and his face in a broad grin. Even a seven-year-old kid could tell that he was gleefully drunk. On his lap sat at least one cheerleader, her midlength blond hair pulled back in the style of the time. I say at least one, because in my memory I have an image of two cheerleaders, one sitting on each of his legs. But I think that this additional cheerleader is probably a hormonally induced embellishment of history.

Regardless of the exact number of cheerleaders on the star player's lap, even at the age of seven, I could feel the powerful message in this image. It's a message that goes back at least to the time of the ancient Greeks. Pursuit of the physical had palpable rewards that were quite distinct from pursuit of the mind. As for the star player, he was able to have his cake and eat it too. The next day, undoubtedly hungover, he ran for over 100 yards and led his team to victory.

If this athlete was a representative example of the student athlete at Nebraska in the sixties, then Nebraska had a good deal to be proud of. He graduated, and after several years as a reserve player in the National Football League, he went on to dental school. When I recently looked up his name on the Web, I found out that he is a practicing dentist in the same town where he played college football.

I don't mean to take the reader for a trip down nostalgia lane to champion the purity of the good old days. There was probably some sleaze that existed on the Nebraska football team of 1963 that went beyond drinking and carousing. But whether significant corruption existed, the level of sleaze was positively benign in comparison to the Nebraska football teams of this current decade. Graduation rates are low (63 percent of those that entered 1990 graduated within six years) and a few of their football players appear to be spending more time with criminal lawyers than university faculty.

As a measure of how much universities overvalue college athletics, coaches of successful major sports teams and their athletics directors routinely earn higher salaries than college presidents. To further sweeten the pot, universities often require their athletes to wear specific brands of shoes and clothes so that their coaches (and the athletics departments) can receive advertising revenue. As a relatively minor example, our basketball players wear Nike shoes and clothes so that our coach can receive several hundred thousand dollars per year from the manufacturer. Adorned with a half a dozen or so tastefully placed

Nike "swooshes," the players are billboards in motion. As a more significant example, the University of North Carolina, our arch-rival in athletics, signed a five year, seven million dollar advertising agreement with Nike in 1997.

What about our athletes in the classroom? How well do they do? And how do schools like mine manage to achieve high graduation rates? I don't get a great number of athletes in my classes, but every once in a while they show up. One year about five baseball players enrolled in a class. Perhaps word had gotten around of my enthusiasm for baseball and there was some hope that I would be kind to them. During the first half of the semester, they hardly attended class and barely did their homework. After the midterm, I wrote a brief note to the athletics department stating that all five of these baseball players were on a trajectory to receive an F in the class. To the athletics department's credit, they must have put the fear of God in these students, because all but one of them soon began attending regularly and worked diligently on their homework (or had a tutor who diligently worked on their homework). They sat in the classroom with grim faces, silently enduring the torture they felt. Only the one who continued to be absent flunked the class. I wasn't too surprised never to see a baseball player in one of my classes again.

I remember the one who flunked quite well because he came to my office twice at the end of the semester. The first time he came to apologize for his poor performance. He said that it was "hard to concentrate on studies when I keep thinking about how I can help the team." The other time he came one hour before the final exam and asked whether "it would be worth my while to study for the final." I told him that studying for one hour was better than not studying at all. He seemed to take this advice to heart, and I showed him a room upstairs where he could study for the remainder of the hour (for the curious, he recorded a 23 percent on the final exam).

But I would be unfair if I did not note that, aside from my brief exposure to the baseball team, my experience with athletes has been largely positive. I am sure that I am not on the athletics department's list of preferred professors. For example, an athlete who did well in one of my classes said to me, "The coaching staff is not happy that I've enrolled in this class." My reputation has apparently served as a useful filter. Athletes that have attended my classes have tended to be serious. In my classes, I have had quite a few women soccer players (our women's soccer team is often ranked in the top ten nationwide) and a couple of male athletes on major sports teams. I have been impressed by these students' ability to juggle what is clearly a very busy practice and game schedule with their studies. Very few of the athletes I have had in my classes harbor any goals of being a professional in their sport, and they often use their athletic scholarships to get as good a college education as can be had.

Athletic scholarships or no athletic scholarships, to be a good student and be on an intercollegiate athletic team in any sport requires the development of a great deal of discipline that is not only useful in school, but in life after school. Being on an athletic team in college seems to be nearly the equivalent of having a full-time job. There are long hours of practice and frequent trips out of town for games. Student athletes, even with help from tutors, are at a distinct disadvantage because they are unable to be present in the classroom during a fair amount of the school year. Out of town trips create conflicts with class deadlines that are sometimes difficult to overcome. It is usually true that even disciplined student athletes end up with grades in my classes that are slightly lower than they would be if they didn't have time conflicts. However, I have had student athletes receive As in my classes and one of them was clearly the best student in a class.

So aside from my brief exposure to the academic abilities of my university's baseball team, I have had little experience with the sleazy side of scholarship athletics. But a friend in another department who teaches an introductory class in his field and is known for his light workload and easy grading policy (and attracts a large number of students in his classes, both athletes and nonathletes), gets regular exposure to the athletes who have no business being in school. He is far more pragmatic than I and has adopted the attitude that if the university has accepted these students, then he will help to find a means to get them through and get their diploma. If I am a bit of a Scrooge to athletes, then he is a bit of a Santa Claus. Despite our different attitudes about the meaning of college education, we get along very well.

Of the athletes my friend has taught, he remarks, "Sure a few are as dumb as a box of rocks, but some of them try hard." Of an athlete who was suspended for a semester for cheating in a class (that was not his class) he says, "I understand why he cheated. He's so stupid that it [cheating] was a matter of survival."

If universities have to accept athletes that are as "dumb as a box of rocks" and have to cheat as a "matter of survival" in order to have the talent to do well in intercollegiate athletics, then there is something severely wrong with our athletics programs. Many private universities and colleges understood long ago that it was not possible to balance serious education and research with scholarship athletics. If it wasn't possible in the past, it certainly isn't possible with so much money at stake now. These schools, including both Ivy League research universities and small liberal arts colleges, often still manage to get their students and alumni excited about their sports teams without athletic scholarship programs.

If you were to talk to a basketball or football coach at a university that is an athletics powerhouse, I am sure that he would tell you that there are many athletes that he does not even bother to recruit because they don't meet academic standards. He might even tell you that there are athletes that he does actively recruit and who he thinks could get by academically that are denied admission. It's not as if university administrations give the athletics departments everything they want. But universities do protect, either purposefully or inadvertently, the eligibility of athletes far more than they should.

For example, Duke, like almost all universities, offers independent study classes. These classes allow a student to work one-on-one with a professor. The professor serves to guide the student in the study of a topic not available in a class or in a research project. Because independent study classes require self-discipline to be successful, they should be reserved for excellent students who are strongly motivated. My own experience has shown me that this form of instruction does not work well with average students, much less poor ones.

Unfortunately, these classes have no accountability except for whatever arrangements are made between student and professor. As a result they are ripe for abuse. A poor student can potentially sign up for an independent study class with a lenient professor, do little work, and receive a high grade. My university currently has no limits on the number of independent study classes that can be taken by a student.

A student must take a total of thirty-four classes to graduate from Duke. For science and engineering majors, the required number of regular classes (not independent study) is typically about thirty. In the nonsciences, the required number of regular classes is typically about twenty. This means that a student can potentially enroll in a baker's dozen of independent study classes and graduate.

Are independent study classes abused? During my second year, I served on a committee that examined this question. For some reason, we were never given complete information to examine the issue in detail. The limited information we did receive suggested that abuse was present, but not rampant. Some athletes and regular students use independent study as an easy way to get through the university. For example, during one year one faculty member known for low expectations enrolled over 20 different students in independent study classes. For these independent study classes, the students were required to write a standard term paper. In typical regular classes, term papers represent about 10 to 30 percent of the class workload. In essence, students were receiving credit for an entire class but did only a small fraction of the work typically required.

The reader might think that it is the professor with low expectations who is principally at fault, not the student. I would agree with this statement, but my university, like all others that I know, does not routinely take to task professors who do not uphold reasonable academic standards. As long as professors show up for their classes and students do not complain about their instruction, usually little is done. So instead of policing the source of the problem, the faculty member, we must awkwardly police the student.

In response to small but persistent abuses like this, the committee that I was on recommended what we thought was a very modest change in the nature of independent study. We recommended unanimously a limit of one independent study class per semester. Such a limitation we thought would not harm excellent disciplined students who wished to work on independent study projects. And it would limit the potential maximum abuse to eight classes over a student's four years. A poor student would still have to enroll in at least twenty-six regular classes (with hopefully reasonably high expectations) to graduate.

For unknown reasons, our recommendation never went anywhere. It was never voted upon by the Faculty Council (the legislative arm of the faculty) and died a quiet death. Undoubtedly, the passage of this change would have had an adverse effect on some of our athletes. Those athletes with poor academic skills would no longer be allowed to enroll in two independent study classes in a semester to keep down their workload. This would also mean that they would have to spend more time competing with our generally academically gifted student population for grades. The chances are that, even though the athlete who was a poor student would preferentially enroll in easy classes, a higher percentage of them would not graduate, or would lose their athletic eligibility during a portion of their four years.

I don't mean to single out Duke. The academic allowances made for athletes are undoubtedly worse at most other places. Across the nation, athletics departments have large staffs devoted to making sure that their scholarship athletes have every opportunity to pass courses and graduate. They give the athletes tutors and inform them which classes and majors are the easiest. For the truly desperate, they find independent study classes taught by faculty sympathetic to athletics or faculty who are just plain pushovers. They are like accountants who find their clients valid tax loopholes. It is embarrassing. Universities are, in essence, giving athletes all of the tools necessary to subvert the objective of providing a student with a quality education. To be fair, it is probably true that most athletes do not take the easy route to a degree available to them.

Athletics officials at universities, when pressed, say that competitive big-time sports are necessary for the health of the entire athletics program. They note that if football and basketball would not earn its millions then universities would not have the funds to offer scholarships in tennis, baseball, soccer, and other sports. They also note that revenues from big-time college sports are used to enhance aspects of the university outside of athletics. Rationalizing sleazy behavior on the basis of "necessary" economics is an old pastime.

But it's clearly much more than economics that drives our passion for big-time college sports. For example, let's go back to the basketball game at Cameron Indoor Stadium. My friend and I are thoroughly enjoying ourselves. There are two minutes left in the game and it has been a seesaw contest. The crowd is exhilarated by the drama. But in the back of their collective minds, there is a certain peacefulness that is palpable in the stadium. Duke, they know, hardly ever loses at home, and there are four future NBA players starting on the Duke team. They are confident of a win and sure enough, in the final two minutes, the team puts on a spurt and the opposition wilts. When the final horn sounds, people rise from their seats. Filing out of the stadium they are content. The opposition played well enough to make it interesting. Many of the fans have the look on their face of someone who has been fed a well-prepared meal.

What is it worth to produce that feeling of emotional well-being on the part of a community? How much corruption is acceptable? This last question is one as old as the Bible. While the problems associated with college athletics are not of biblical proportions, let's do the math for a typical topflight basketball team. On the team, there are about thirteen scholarship athletes. Probably three or four of them have good academic skills and have fair to excellent grades. Another three to five have borderline academic skills and with enough tutoring and hard work can do passably well in classes. Sure they wouldn't be at the university if they couldn't play outstanding basketball, but the truth is that private universities sometimes accept children of wealthy alumni with similar academic skills.

Finally, there are the three to five players who absolutely have no business being in a university. Some are star players and some never live up to the athletic expectations placed upon them. Through tutoring, enrollment in carefully selected classes (some of which a scarecrow could pass), and independent study classes a university will keep them enrolled and hopefully graduate two or three of them. So all in all, in order for a university to be highly competitive on the basketball court, it has to completely abandon academic standards for three to five out of thousands of undergraduates and seriously bend academic standards for another three to five. These are obviously small numbers. And when one looks at an entire athletics program, there are likely at most 100 "students-in-name-only" that a university accepts and manages to keep enrolled so that it can be highly competitive in most sports. Is 100 too large a number? Is fifty too large a number? Is ten too large a number? For me, the answer is yes to all of these questions. But I know that most people involved with universities are very tolerant of the academic compromises that are made to accommodate competitive athletic programs.

Research
and Graduate Education

6

Heart and Soul

It was February of 1993 and my department was holding its monthly faculty meeting. We were a small department of twelve faculty members and ten of us were seated in our conference room. The other two would have been there but were out of town. For most of our faculty meetings, this level of interest and attendance was not common. But this day's meeting was about money and research. We were collectively deciding whom to admit for graduate study the following fall.

We sat around four large blonde oak laminate tables that were pushed together to form one very large surface. The conference room was recently remodeled along with the rest of the building and painted a bright white. Looking out the window you could see the Law School office where Richard Nixon, a Duke alumnus, was rumored to have studied in the 1930s. As a department we got along well. The politics sometimes got rough, as they typically do at universities, but there was little paranoia or backstabbing. If the ghost of Richard Nixon existed on campus, it must have only rarely entered our building.

There were fifty applicants for graduate study. Before the meeting, each of us had rated each applicant on a scale of one to ten (ten being the highest rating). A secretary had tabulated all of our scores and computed an average score for each student.

Our ratings were based largely on grade point average (GPA) and test scores on the Graduate Records Exam (GRE). Recommendations were potentially important but they were generally discounted. The problem was that it was difficult to find valuable assessments of students. Most students were able to find at least three faculty members at their undergraduate institution who were willing to write uncritically and deliver pro forma glowingly positive reference letters. I have had applicants with C plus average grades with uniform recommendations attesting to the outstanding academic abilities of the student.

But C plus average grades in our applicant pool in 1993 were rare. We had hired five new faculty members in our department from 1988 to 1992. As a result, we had developed a reputation as a good place at which to receive a Ph.D. We were a department whose research activity was on the upswing. And so was the quality of our applicants. On paper most of our applicants had B plus to A average grades. The average GPA at most institutions in 1993 ranged from 3.3 to 3.5 so these were average to excellent undergraduate students. Their GRE scores were good to excellent. Our applicant pool consisted of a mixture of outstanding students who were using us as a backup school for graduate study and good students for whom we were their first choice.

For our fifty applicants, five teaching assistant (TA) positions were available. TAs were usually our sole means of attracting new students. While almost all of us had active grants for our research, we usually used grants to fund students who already had taken enough class work to be useful as research assistants. The TAs received a modest salary ($8500 over nine months) and free tuition. Since graduate school tuition was about $20,000 a year, we expected that only those who were offered a TA would enroll. Typically, about half of those to whom we offer a TA choose to study elsewhere. So we overbook and initially offer one to two more TAs than we have available. Our job at this meeting was to create a list of applicants to offer TAs and applicants to place on our wait list for TAs.

In order to make this decision, we used the average scores from the faculty ratings as a starting point. The names of the top eighteen ranked candidates were written on the chalkboard in order of their score. Almost all of the students had indicated that they wished to study with a particular faculty member. And almost every faculty member attending had a vested interest in at least one prospective student. To start the meeting, the Director of Graduate Studies asked the question, "Does anyone want to make the case for changing the ranking of a student?"

I certainly did. Two students who wanted to study with me were listed on the chalkboard. But that year my prospective students weren't highly ranked. One was ranked number eleven. He came from a small, somewhat obscure, midwestern liberal arts school with a reputation as a very good institution that took education very seriously. He had received As in all but one of his undergraduate courses. Because his GRE scores were just average, he had a rating of only 7.8. The other was ranked number fifteen. She came from a northeastern institution well known as a party school (one "insider" college guide listed it as the best campus for beer) and had average grades and high GRE scores. She had a rating of 7.7. For all intents and purposes, their ratings were identical. But the ratings were based strictly on their files, and by 1993 I knew that you couldn't make judgments on grades and test scores alone.

So I made phone calls. I talked to the applicants. I talked to their former employers and former teachers. My principal goal in these conversations was to determine how driven the student was to succeed. Strength of purpose is, to my mind, a key element in whether someone is successful in research. That and creativity are the difference between an average or poor graduate student and an outstanding one. I couldn't judge creativity on the phone. I could, however, ask questions about these students' past performances and their reasons for going to graduate school.

On the basis of these conversations, I felt fairly confident that number 11 would be a valuable graduate student. Number 15, however, would likely go through the motions and give me heartburn. If the ratings were to remain unchanged, both would be put on the wait list for TAs. But I didn't want that to happen because I knew that some other school would offer number 11 a TA at about the same time we would put him on the wait list (all universities make their decisions about graduate students at about the same time in February). Unless I could get him raised up to number seven on the ranking and off the wait list, I would likely lose him.

So at this meeting, I tried to lobby for candidate number 11. Moving him up four notches was probably impossible. We tended to stick with our quantitative rankings and were not generally moved by qualitative arguments. Still, I tried my lawyerly best. "His grades are the highest of the entire applicant pool," I said. "He comes from a school well known for producing quality students in the earth sciences, some of whom have gone on to prominent academic positions. He has been a research assistant for three summers at the Environmental Protection

Agency. I've talked to his employers there and they say that he is serious, hardworking, and incredibly efficient. One of them said that 'I would be a fool' if I didn't accept him.

"The only reason this student is not in the top five is because of his GREs and they aren't that bad. He has had more research experience than any other candidate and is a proven performer. I would take a strongly driven achievement oriented student like this one over one with 1450 GREs any day of the week. I think he should be moved up."

This little speech was surprisingly effective. Most everyone agreed that we probably overemphasize GRE scores. Perhaps this was because the year previous we had a couple of students that we accepted with high GRE scores who had turned out to be duds. Other people lobbied to move other candidates up, but were not successful that year. I managed to get my candidate moved up to number eight. Then I pushed to overbook by three this year so that he could be offered a TA and not be put on the wait list. "If worse comes to worse and five people ahead of him accept, I'll pay for him out of a grant." Everyone was agreeable to this idea.

Now all I had to do was convince that student that my university was the best place for him to study out of the three schools where he had been accepted with financial aid. I telephoned, and offered to cover his expenses to visit the university (the money for wooing potential graduate students of high quality came from the Graduate School). In striking contrast to other universities where he was being wooed, the weather was beautiful when he arrived for his visit. I put him up in a nice hotel. I fed him good food. I showed him all of the good things about my university. I was not very surprised when, in March, I got a call from him stating that he had decided to come to study with me.

In hindsight all of this maneuvering seems obsessive. But my efforts to attract a quality graduate student were not out of the ordinary. As a consequence of the Golden Age, every world class university or university that aspires to be world class is obsessed with the recruitment, education, and research of their graduate students. We are obsessive because graduate recruitment and education, more than any other investment, pays intellectual dividends. Through their hard work in teaching and research for low pay (about the same hourly wage as a Burger King salesclerk), they are dollar for dollar far more important than the faculty. Graduate students are the heart and soul of the modern research university.

In a nutshell, if your graduate students are successful in research then your university is successful. Graduate student dissertations constitute the bulk of research at any university. As in any creative enterprise, much conventional research is done in order to produce the occasional piece of cutting-edge and pathbreaking work. A university's reputation in the sciences is largely dependent upon that occasional Ph.D. dissertation that is widely accepted as an intellectual tour de force in the academic world and perhaps in newspapers or major corporations. The supervising faculty member is partly responsible for such highly successful research, but it is the graduate student who does the actual work.

It's not only the path breakers that are important. Every Ph.D. dissertation, and its associated articles, books, and sometimes patents collectively add up to elevate a university's reputation. Students who produce good, but not necessarily cutting-edge work often take academic positions in prominent institutions. When they do well at those institutions, the reputation of their alma mater is enhanced. The overall reputation of any department in the sciences is almost always directly related to the success of its Ph.D. graduates.

It wasn't always this way. Before the Golden Age of the American research university, graduate programs were much smaller and many faculty in the sciences performed much of their own work. The availability of easy federal grant money, particularly after the Russians showed their technological prowess (and our relative lack thereof) with the launching of Sputnik, changed all of that. Money was available for far more research than faculty members could perform themselves. They needed a cheap, smart, hardworking labor force to do the research. The Golden Age of the American research university spawned the Golden Age of the graduate student.

Under the established procedure of the Golden Age, the principal roles of a faculty member in the sciences are to raise the money to facilitate graduate student research and to review graduate student work and to advise periodically. So when a faculty member is said to be an active researcher in the sciences, it tends to mean that he or she is actively supervising and raising money for several graduate students. It is a continual challenge to raise money, especially in these days of increased competition. Grants for research are usually about two years in length. If you have a few active grants, it usually means that during every quarterly period there are a couple of status reports to fill out, a couple of new grant proposals to write to ensure continued funding, and of course a research paper or two to write because if you don't publish your work, you can kiss future funding good-bye.

For an active researcher, training of graduate students is more than a full-time job. It's like running a company, and the financial pressures to keep things going are constant. At one time, I had seven graduate students working for me. In the days when grant money was plentiful, this would not have been at all unusual. But with the end of the Golden Age, making sure that all of these seven students had salaries and the money to do their research was trying.

There were times when I would wake up in the middle of the night worrying that if I did not get a certain grant funded in the next three months, I would have to tell a graduate student I had no money for him. My financial pressures were similar to that of my father, who had a small construction business. He too woke frequently in the middle of the night because he was worried about how he was going to pay *his* workers. I was in the same boat, except that I was not a businessman. I was a scholar. But the rules of the Golden Age had turned me into a manager of scholars. My university expected this evolution in my development and would evaluate me for tenure largely on my ability to make this transition successfully.

It's no wonder that under these rules of behavior and a university's financial dependence upon them, teaching of undergraduate students gets short shrift in the American research university. Undergraduates will continue to come and pay their money regardless of the level of commitment to undergraduate instruction shown by the faculty. Conducting substandard research, however, will result in no future grant funding. With the current high level of competition for grants, conducting very good research on existing grants doesn't even cut it. You have to produce excellent to outstanding work to have a decent chance at keeping the money coming. So there is strong tendency to focus on research because the standards for continued revenue for research are higher than the standards for continued revenue from undergraduate instruction.

When I came to Duke, attracting excellent graduate students like the one I found in 1993 was difficult. In contrast to 1993, our graduate program in 1990 was of only fair quality. The faculty hires that would give us the critical mass necessary for respectability were still being made. Some of the students were good to excellent, but they were the exception to the rule. These students had usually chosen my department for logistical rather than intellectual reasons. They had spouses or significant others with jobs in the area. Or they were from the South and my university was as far north as they were willing to move.

While finding good graduate students was difficult in 1990 at my university, it was becoming more difficult in my field of study even in highly respected departments. Over the years, earth science departments had tended to require less of undergraduates in an effort to keep class enrollments high. Also, quite a few applicants in my field of study had followed the fashionable trend of choosing their undergraduate major in environmental studies. Science requirements for this major are weak virtually everywhere. As a result, students who applied for graduate study in the earth sciences across the country often had successfully managed to avoid introductory calculus, physics, or chemistry as undergraduates. The problem with watering down the undergraduate curriculum was that a student needs to know basic calculus, physics, and chemistry to do research in the earth or environmental sciences.

I didn't understand these limitations when I first came to Duke. I was very naive. I thought that evaluation of applicants could be done strictly on the basis of their GPA, GRE scores, and recommendations. I thought that students without drive and dedication wouldn't even think of applying to graduate school. I didn't know that a student with a 3.0 GPA in 1990 was typically a student in the lower third of his or her graduating class. I didn't know that at many universities undergraduate classes were so easy that a student didn't have to work and could still get good grades. And as a result, my early efforts at picking and training graduate students were not always successful.

Some of the problem was not my own doing. There were low spots in our graduate program in 1990. Sometimes our standards dropped to levels that were beyond belief. For example, consider a graduate student of mine who was funded with money from his nation's government. For reasons that have never been explained to me, he was accepted by my department's Director of Graduate Studies without any input from myself or anyone else. This student could barely speak or understand a word of English, and I didn't know he existed until he arrived on campus to study.

The student knocked on the door of my office three weeks into the semester (he was three weeks late because he had had problems with his visa). After twenty minutes of painful conversation, he was finally able to communicate to me that he was a new student who had arrived late and had come to study with me. I was so incredulous about this situation I decided that what he managed to communicate to me must be a joke. This was a warped hoax, I thought. Someone had managed to get a foreign graduate student to pretend he was new and couldn't speak a word of English.

I spent the next thirty minutes with him barely able to communicate anything, but by that time I was so convinced I was the butt of a joke that I was smiling. I marveled at how good an actor this student was. Maybe he wasn't a student after all, but an actor friend of another faculty member. Before he left, we shook hands and I told him that I hoped to see him again soon. I thought that if I ever saw him outside of my office, I would congratulate him on his acting skills.

Later that day, after no one called to slyly ask whether I had received any visitors, I decided to see whether this person's story could possibly be true. I went into the department's graduate student record files. To my surprise, I found his file and an accepted graduate application complete with a photo. In the portion of the application where you state the specialty you wish to study, my area of specialization was written down very neatly and clearly. I looked at his file. He had received an undergraduate degree from the most respected university in his country and had good grades. So far, so good. Then I looked at his test scores in English writing and comprehension (every potential student from a non-English speaking country must take a standard test in English in order to be considered for admission). I then called the graduate admissions office to find out just what these test scores meant. "Those scores are below our cutoff," the woman in the graduate admissions office stated.

"How much below our cutoff?" I asked.

"Well it's not a very good score," she said. It turned out that he was not even close to the minimum acceptable proficiency in English to enter graduate school. When I asked how he had managed to get admitted, I was told that someone must have made a special case for him.

As for the student, he tried, but never developed a reasonable command of English. I sheltered him for a year by having him take classes that were heavily steeped in mathematics. But in the second year he was expelled for poor grades.

My other early graduate students possessed no language problems. They were all American born. On paper, none of them were outstanding applicants, but they all had reasonable to good undergraduate records. They all had excellent recommendations. But they had not been stretched in their undergraduate years and their efforts in making the necessary change to dedicate themselves to their studies produced mixed results.

It was not just independent research that proved to be a stumbling block. The classes tended to be harder and as a result they could no longer coast and get by as they did as undergraduates. Some had come to graduate school partly because the job market was poor and graduate school provided the potential to extend their college experience.

Also, they knew that my field of study promised many jobs. So some viewed their studies partly as vocational training.

Once, I was quizzing a student about what he had done on his research during the week. It turned out he had done virtually nothing and admitted it openly. I expressed my displeasure over this lack of progress. Instead of responding that he would do better next week he said, "Stuart, you just don't understand. If I had my druthers I would sit on the couch and watch football and movies on the TV all day long. Your other students are the same way. You're just going to have to get used to it."

He was not correct in his generalization of his attitude to all of my graduate students. It was, however, probably the most honest thing a student has ever said to me. In the workplace, you could just fire someone like this on the spot and move on. But graduate students require a good year of course work before they can perform any real research. I had a grant and needed to show that I was producing significant results. I had to ride this horse wherever it would take me.

It could have been worse. Once, a friend was in the Sierra Nevada Mountains of California helping a student core rocks to obtain samples for laboratory analysis. In the middle of their fieldwork, they had worn out all of their drill bits. The student, who knew how many samples would be collected beforehand, had not ordered nearly enough drill bits. Standing in front of their field vehicle, my friend asked him if he knew that they were running short.

"Yeah, I noticed several days ago that we were running out," his student said.

"So why didn't you order more drill bits then so that we would not be stranded in the field and have to wait two days before we could get more work done?" my friend asked, with more than a little anger in his voice.

The student looked at my friend with indignation. "Why don't I take my dick out, put it on the hood of the truck and you can whack it off with an axe?" he asked in what my friend hoped was a rhetorical question. This student had a 3.6 GPA from a highly regarded institution and excellent recommendations, one of which extolled his positive attitude and work ethic. Perhaps the recommendation was an honest one and reflected a comparison with other students. Somehow, I don't think so.

My own biggest problem was that there was a significant gap between what I needed from my first batch of graduate students and what they could provide. They were by and large nice people and most of them tried hard. But they could only do so much. Because of this,

I was not able to do what I said I was going to do in my grant proposals.

By the spring of 1993, I was beginning to turn the corner with my first batch of graduate students. I knew that they would all graduate with master of science degrees within the next year and not continue for their Ph.D. My next generation of graduate students was more talented. I had learned that I needed to treat a graduate application as thoroughly as a high quality employer treats a job application. I no longer accepted students on the basis of respectable grades, recommendations, and standardized test scores. Before I accepted a student, I went out and got the information I needed to determine whether that student had the heart, conviction, and talent to do good work in graduate school.

As a result, over the next four years working with graduate students became a very rewarding experience. My students were, by and large, smart and productive. When you have good students, they not only learn from you, but you learn from them. I guide the basic direction of work, but they often come up with great ideas on their own.

For example, the student for whom I finessed a teaching assistantship in 1993 was a joy to work with. He was strongly driven to do a good job on his research. One day he got the idea that it would be worthwhile to analyze some of the water samples that he collected for his research for a certain chemical constituent. But we did not have the laboratory facilities for such analyses. Undeterred, he drove his samples to a university 300 miles away and used their laboratory for two weeks. These analyses became an important component of that student's thesis.

The downside of my circumspect approach to bringing in graduate students was that I accepted far fewer students than I did in the beginning and had a much smaller graduate research program. To partially make up for this, I hired postdoctoral students (students who have already earned their Ph.D.). Postdocs earn a much higher salary than graduate students and hence the overall size of my research group was much smaller. While the volume of research produced by my students and postdocs dropped, the quality improved markedly. And because I had fewer students, I started to do some research myself, which is why I became involved in research in the first place. All in all, I was much happier operating in this mode.

I also saw this change as a positive one for my teaching. I was in closer touch to research day-to-day, and I was learning more than I did during my brief tenure as a manager of a larger program. Unlike those faculty who were operating in the classic mode of the Golden Age, when I taught I did so from firsthand experience that grew with every year. Over the long haul, my classes would be able to maintain their freshness.

Finally, I think that this change to a smaller research operation is (at the risk of trumpeting my own horn far too much) a good model for the post-Golden Age era. Certainly, the model of science faculty members managing large herds of graduate students is no longer practical. Even if one can find the research funds and students to do this, the end of the Golden Age has meant that in most fields there aren't jobs for most graduate students after they graduate. Large programs are a remnant from a time when there was a great demand for Ph.D.s in academia or industrial and government research laboratories across the sciences. The end of growth of federal funds for research and the end of the era of the well-funded corporate research laboratory have caused a glut of Ph.D.s in the marketplace.

The situation has led to gallows humor among students in many fields. For example, a joke around the Duke physics department is that the job market for Ph.D.s in any given year is "average." It's much worse than the year before, but it's much better than it will be next year. There is no doubt that we need to downsize most of our graduate programs, including those in the humanities. For faculty members who continue to receive large amounts of research money, postdocs and research technicians can often replace graduate students. Graduate programs are important, but they should no longer dominate the research and teaching agenda of the American research university.

7

Grants or Goodbye

I was driving from Durham, North Carolina, to Washington, D.C. It was a drive of 260 miles or so through the pine forests of North Carolina and Virginia on interstate highways. I had made this drive once before with my family to visit my wife's cousin. Back then, we followed the speed limit and stopped for lunch along the way. All told it was a five and a half hour drive.

But I didn't have five and a half hours for this trip. I had left Durham after dropping my daughter off at school at 8:30 A.M. and had promised to come back to Durham and pick her up at a friend's house at 5:00 P.M. So all told I had eight and a half hours to drive up to D.C., do my errand, and drive back. I could follow this timetable as long as I drove twenty miles over the speed limit, which is what I was doing.

It was June 1, 1991, and my first academic year at my university had ended. So according to the public's general view of the life of a university professor, I should have been relaxing and enjoying my summer vacation. But I wasn't relaxing. My adrenaline level was high. June 1 was the deadline to hand in proposals for the National Science Foundation (NSF). I had finished a proposal the day before, too late to send by overnight mail. If I missed this deadline, I would have to wait another six months for my proposal to be considered for funding. So I came up with the rather crazy idea of driving the proposal to D.C. by myself.

When I took my job at my university, I was aware that raising money through grants was important. At a personal level, I knew it was financially lucrative. Included in the budget of most grants is a month or two of summer salary. If I wanted a summer paycheck, I needed to get a grant funded. But it wasn't until the first meeting of untenured faculty in October (discussed in detail in another chapter), that I understood the external pressure to obtain grants.

At that meeting, I had asked the chair of the Appointment, Promotion, and Tenure (APT) Committee a question. "Suppose, hypothetically," I said, "in the next seven years I work on problems that don't require research money or graduate students. I publish a lot of research papers, many of which are highly regarded, teach well, but don't raise any grant money. Will I get tenure?"

This was, I knew, a somewhat pretentious question. But I wasn't being difficult for the hell of it. Part of me wanted to follow this hypothetical path. There are numerous interesting problems in hydrology that don't require graduate students or a lot of money. It seemed to me a romantic ideal to avoid grant writing and concentrate on research and teaching. If I could stand the lack of summer salary, why not? After all, it was a model with an extensive precedent. It was practiced widely prior to the Golden Age of the American university.

Part of me wanted to be such a throwback. At heart, I am a purist. In baseball, I can't stand artificial turf. I find the idea of a designated hitter absurd. I'd much rather watch a day game than one under the lights. If I had been alive in 1919, I would have likely been against the livening of the baseball to allow for more home runs. So the question I asked the chair of the APT was an honest one. In response, he did not flinch.

"Receiving grants from peer-reviewed proposals is an indication that your peers value your research," he replied. "I would think that it would not be possible to publish highly regarded science articles and not receive grants." In other words, I wasn't just being told to publish or perish. I was also being told to get grants or say good-bye.

No one had been so up front to me about the university's expectation that I obtain grants until then. Still, I should not have been too surprised. It was fairly old news. The Golden Age of the American research university was fueled by federal grant money. And while the Golden Age had ended, the need for federal grants was still very strong.

Before World War II, the American university was a centuries-old quaint and charming haven for scholarship and education. After World War II (and particularly after the Korean War) it quickly developed into a supercharged intellectual powerhouse heavily dependent upon federal grants for its economic health. How had this happened? Why was I being told that without grant money I would not receive tenure? It all started in 1944.

In 1944, the Manhattan Project (the project that built the atomic bomb) and other science-based projects with military aims were positive proof that massive federal funding of scientists could be of great use to the country. To give you an idea of the level of federal commitment to military oriented science, roughly two billion dollars was spent on the Manhattan Project during World War II. President Roosevelt, with an eye toward the future, requested that Vannevar Bush write a document detailing the potential role of public funding of science after the war.

Dr. Bush, a highly regarded electrical engineering professor at MIT and one of the fathers of the computer age, had taken a leave of absence to be in charge of army research during the war years. In response to President Roosevelt's request, he prepared an amazing document that championed the value of science to the nation. Entitled "The Endless Frontier," this document, which is still extensively cited, laid out the map for the funding of scientific research after World War II.

Professor Bush's principal argument was that federal funding of science was necessary for the improvement of the nation. Science had to be broadly funded because it was impossible to predict what aspects would prove of value to society. A major aspect of science that should be funded was fundamental and theoretical work. Central to Professor Bush's argument was the notion that the public could not understand the nature of science, but that they should accept its implicit value. Scientists should be given a pool of money and scientists should decide amongst themselves who should receive that money.

Even Vannevar Bush could not likely foresee how rapidly the pool of money would grow. In 1944, he suggested that about thirty million dollars for peacetime federal funding of all scientific research should be distributed across federal agencies and universities. By 1997, funding of health-related research in universities alone had grown to seven billion dollars.

The initiation of large-scale federal funding of science provided a great opportunity both financially and intellectually for universities across the nation. But not every college or university wanted to participate. It was generally acknowledged that federally funded research would change the face of the university. The question was whether that change would be beneficial. Some feared that federal grant money would make universities beholden to politicians. Others feared that fields of study for which federal funds were not readily available would atrophy.

In response to this new source of money, my own university was cautious and initially chose a mixed approach. In the 1940s and 1950s, it decided to chase after federal dollars in the medical and life sciences. Other sectors of the sciences and engineering continued to focus on undergraduate instruction. It wasn't until the 1980s that Duke decided to broaden its aggressive search for federal dollars across all of the sciences and engineering.

But in the 1950s, many universities saw an advantage in building up all of the sciences and engineering with an eye toward obtaining federal grant money. No single institution did this better than Stanford University. And no one person was more attuned to using federal grants to build a university than its dean of engineering and eventual provost, Frederick Terman. Dr. Terman, trained as an electrical engineer, was an aggressive man with insight and boundless energy. When he began as dean, Stanford University was considered a good private regional school. When he retired as provost, it was arguably the best university for research in the nation. Vannevar Bush prepared the blueprint for the Golden Age of the American university. Frederick Terman, who earned his Ph.D. in Bush's department at MIT, was a master builder who knew exactly how to use that blueprint. He was the model leader for the American research university in its Golden Age.

Terman knew that in order to take full advantage of federal funding, his university would have to retool and emphasize the sciences and engineering. Before federal funding became widely available, science and engineering departments at Stanford and most everywhere else were relatively small. American universities traditionally emphasized the humanities and many were slow to recognize the rise of science that took place in Europe. The decisions made by Terman at Stanford and mimicked by others elsewhere changed all of that.

Following the adage that it takes money to make money, he went on a hiring spree to obtain the best and brightest. The ecologist E. O. Wilson, in his autobiography *Naturalist*, tells a story about Terman. When Wilson was a non-tenure track assistant professor at Harvard, he received a letter from a Stanford administrator offering him a tenured professorship. The letter came as a complete surprise since he had not applied for the position. Soon afterward, Terman, with another man in tow, made an unannounced visit to Wilson's Harvard office. Terman introduced himself as Stanford's provost. Wilson turned to the man accompanying Terman and asked, "Are you from Stanford, too?"

"Yes," Terman answered for the man. "He's the president."

Wilson, while impressed with this level of wooing, ultimately said no to Stanford, and chose to take a tenured position at Harvard.

But many other scientists and engineers did not say no. Terman was an excellent salesman. He also knew what merchandise to buy. For example, he revamped his chemistry department by raiding prominent faculty from other universities. Three of these chemists went on to win Nobel prizes. The money to create a world-class chemistry department was not all internal. A good portion came from a wealthy donor who provided the funds for a new chemistry building. Terman's success at fund raising and recruiting was university wide, and his efforts to build the sciences and engineering paid big dividends intellectually and economically. His hiring efforts produced many members of the prestigious National Academy of Sciences. As federal funding of sciences and engineering grew exponentially, so did the revenue of his university.

For those universities like Stanford (and to a lesser degree Duke) that actively participated in the exponential growth of federal funding of science, buildings were constructed and administration and faculty size increased. Faculty salaries rose to a level approaching those of conservative Fortune 500 companies, particularly for those in the sciences and engineering. With an increase in faculty size came a decrease in teaching load. The increase in faculty size, coupled with the ready availability of research funding, also created a climate ripe for training new Ph.D. students. Once graduated, they took advantage of newly created positions in growing universities across the nation and frequently found jobs as professors. By 1990, federal research grants typically accounted for 30 to 60 percent of revenue received for academic programs at major universities.

On the plus side, this funding produced the greatest output of science ever seen by any country. The money generated by grants in the sciences was also partly responsible (in addition to revenue increases from tuition) for the aggrandizement of the entire university. The spillover in funds helped social sciences and humanities departments grow as well. Salary increases for faculty members in these areas were significant, as were decreases in teaching load. These professors also began to receive grants from federal agencies, albeit of smaller magnitude than those in the sciences and engineering, and spent more time on research.

But there were some trade-offs. Many faculty members were so attuned to research that their attention to undergraduate education deteriorated. Also, the infrastructure built through the munificence of the federal government was costly to maintain. Post World War II federal funding had profoundly changed the nature of universities. It was somewhat like comparing the family physician of yesteryear to the HMO of today. The university was now a big business with critical day-to-day concerns about cash flow. My university was telling me to get grants or say good-bye in 1990 because grants were now a staple of the university revenue stream. The growth that created my position was largely the result of grant money, and the expectation was that I would justify my position by getting even more grant money.

After the meeting when I asked the question of the chair of APT, I knew I had better get cracking on getting grants. I had no experience writing proposals, but I had reviewed quite a few in my former position in the government. I spent time rereading them, studying the "art" of a good proposal.

It would be a good idea, I thought, to first look for small pots of money designed for starting faculty. I wrote a small grant proposal ($10,000) to the State of North Carolina to purchase equipment for my hydrology laboratory, and within a few months, this grant was approved. But before I received the money, the state program that approved the funding was eliminated. I started to write another small grant proposal ($18,000) to the American Chemical Society. When my department chair heard that I was doing this, he said, "Stuart, why are you bothering with such small amounts? When are you going to go after the big money?" I thought about his view and decided that he was probably right. Come tenure time, successful grants for small sums would matter little. So I dumped the American Chemical Society proposal and began to write a new one to the National Science Foundation (NSF).

In my field of study and for most of those in the nonmedical sciences, NSF is the preeminent funding source for research. In that year, it distributed about two billion dollars to universities. For my first NSF proposal, I decided to analyze data that I had collected in my previous job and propose to extend the work. I began to write this proposal a month before the NSF deadline, thinking that it would take me at most two weeks to pound it out. But two weeks came and went with me working day and night, and I wasn't close to being finished. So I started working even longer hours. Finally, after three weeks, it looked like I was making real progress. I handed in a draft of my proposal to the grants accounting office at the university for approval three days before the deadline. I continued to hone the manuscript until the final day before the deadline.

At 4:00 P.M. on May 31, I finished the proposal and then made a stupid decision. Instead of going to a copy shop to make the required twenty copies of my proposal (NSF is run by scientists, but it is a classic government bureaucracy, hence the need for twenty copies), I decided to use the departmental copy machine. Like almost all copy machines at my university, it was a no-name brand prone to misfeeds and failure. It took me two hours to make the copies. But I still wasn't too worried. In California, where I had lived previously, Federal Express accepted packages for next day delivery until 8:00 P.M. But when I called their office in Durham at 6:00 P.M., they had already closed. Other overnight mail services were closed as well. I still hadn't adapted to life in a small city.

I was upset about Federal Express being closed, but mostly I was upset at myself for my poor planning and decision making. My wife was out of town for the week. I picked up my daughter at a friend's house and went home and cooked dinner for both of us. I thought about what could be done and decided that I should drive the proposal to NSF.

The next day, in my car, I felt somewhat better. I was making good time. In two hours, I neared the city of Richmond, capital of Virginia, and former capital of the Confederacy. Its famous native sons included Edgar Allen Poe and Arthur Ashe. It looked like a city of reasonable size with a well-defined downtown. If the South had won the Civil War, I thought, I would be submitting my proposal here and my drive would be over. It was the first time in my life that I wished that the Confederacy had been victorious.

I continued for another two hours until, after fumbling around the streets of Washington, D.C., for awhile, I double-parked in front of the headquarters of NSF. At the time, NSF was located in the heart of D.C., housed in a nondescript brick building constructed in the fifties or sixties. There are many grand buildings in Washington, D.C., but this wasn't one of them.

I walked into the building with a box holding the twenty copies of the proposal and asked a clerk where I should drop it off. He told me where the mail room was located. Walking into the mail room, all of my energy vanished. The sight was depressing and overwhelming. I was in a large, dark gray room devoid of any people, filled with stacks of proposals that had arrived that day. The piles formed hills some six feet high. I placed my proposal near the top of one of them and walked back to my car.

Driving home, I thought that the mounds of proposals that I had seen looked a little like the pyramids of Giza and it seemed to me that this proposal process was a giant pyramid scheme. Many proposals were written, but very few received funding. What were the chances of mine being picked from all of those proposals? I felt like I had completely wasted my time.

I shouldn't have been so despondent. Five months later, I found out that the odds of my receiving funding from NSF's earth science program were (for that year) 22 out of 100. While at face value the probability of winning this grant money was far worse than winning a typical bet in Las Vegas, the odds weren't too bad. I knew from reviewing other people's proposals that many were poorly conceived and written. Others naively assumed that all you had to have was a good idea and that inclusion of preliminary results was not necessary. Probably less than about one half of the proposals in those piles in the mail room were legitimate contenders. My odds, given that my proposal was carefully thought out and presented a solid case for extending already significant results, were around one in two.

So I shouldn't have been in such a state of shock when a man from NSF called announcing that my proposal had been accepted for funding. I was to receive about $250,000 over two years. I felt like I had won the lottery.

No matter how you cut it, $250,000 is a lot of money. It is very unusual for a young (or even a new and not-so-young) faculty member in the earth sciences to receive this level of funding. It was so unusual that one month later the same person from NSF called to say that they were reconsidering my budget. Could I do the proposed work if the budget was cut by 40 percent? he asked. I felt instant dread. I didn't say anything for what seemed like an eternity. Then I recovered. I explained why all of the money in the budget was necessary, and must have been convincing because I never heard from NSF about this matter again.

So young or not, why would I need so much money? First of all, about 30 percent of this money went directly to my university to use in any way it saw fit. This 30 percent, called overhead, is a university installed tax on all proposals. It helps pay for building maintenance and faculty salaries among other things. So in terms of real dollars, I had about $175,000 to spend over two years.

Most of that money went to support graduate students for the two years. Typically, graduate students in the sciences receive a stipend and free tuition to take classes and perform research. In 1991, we paid our graduate students a stipend of roughly one thousand dollars per month (barely enough to live on, but at least they didn't starve on this salary) and tuition was about eighteen thousand dollars per student. So the cost of funding a graduate student was roughly $30,000 per year. I needed two students to work on this project so $120,000 went to fund these students.

That left about $55,000. Some of that paid for my salary during the summer, when I did much of this research. For NSF grants, each grant is typically only allowed to pay for one month of summer salary per year. The remainder of the money, about $45,000 of the original $250,000, was used for the actual work. It allowed me to buy scientific instruments and to make measurements. For this grant, the major nonstudent-related expense was the cost of drilling a fairly deep well (about 600 feet deep). The second largest nonstudent-related expense was for travel ($75 dollars per day for lodging and food including tax) to perform measurements at the well for a four-week period over the two years of my grant.

It should be noted that if I were to have written this grant three years later the cost of this work would have been considerably less. Student-related expenses would have dropped dramatically. The end of the Golden Age of science funding meant that agencies like NSF became very cost conscious. They looked askance at the practice of paying for graduate student tuition through grants. As a result, tuition was no longer required of graduate students as of 1994 at the university. A modest yearly $3000 registration fee was installed in its place. So had I written this proposal in 1994, the budget (with reduced student expenses and resulting reduced total overhead) would have been about $90,000 less. This amount is close to the 40 percent reduction in budget that NSF originally requested. Apparently, NSF had called me about reducing my budget three years too soon.

When I first received notice that the NSF grant had been funded, I told everyone that I knew and loved, including my wife, my daughter, my mother, and my in-laws. I was ecstatic. It was like having a job as a salesman and getting your first big sale. They all knew the pressure I felt to obtain grants and were pleased that I was showing signs of success at my new job. When I told my colleagues at work, there was a positive shift in their opinion of me. By getting a big grant, I had shown that I was a member of the club.

Over the next six years I would write a total of twenty-five proposals to federal and state agencies. To give you a feel for the nature of these proposals, here are some sample titles:

▼   Temporal Changes in Permeability Induced by Seismicity
▼   Determination of Air Permeability of Soils from the Response of the Unsaturated Zone to Atmospheric Pressure
▼   Using Groundwater Levels as Strainmeters, Anza Seismic Gap, California
▼   Permeability Structure in the Presence of Sparse Data: The MADE Data Set

Clearly, these titles were not written with the lay public in mind. And I know from experience that talking about these research topics to nonscientists causes instant drowsiness in the listener. However, not one of my proposed research projects would have been a likely candidate for the late Senator Proxmire's monthly Golden Fleece award for the most wasteful use of federal research dollars. For example, consider the second yawn-inspiring title on the list. Determination of air permeability is an important component of cleaning up soils that have been contaminated with toxic waste. All of these proposals had some societal value.

It took a combined twenty months of full-time labor to write the twenty-five proposals. So over my first seven years as a professor, I spent roughly a quarter of my time trying to get money. Six of the twenty-five proposals received funding. To be honest, not all of my proposals were as carefully written as my first one. And many of them were written to agencies where the odds of receiving money were much lower than at NSF in 1991. Having a successful proposal percentage of 24 percent for my first seven years was, on the whole, doing fairly well.

However, I don't expect that my future success rates will continue to be so high. The end of the Golden Age has caused a significant drop in the acceptance rates of proposals. The cause of this drop is not due to a decline in the pool of money: The pool of funding has been increasing at a level equal to the rate of inflation. It's just that during the Golden Age we produced too many Ph.D. graduates. Some, like me, have taken academic jobs. Others have taken postdocs. We are all applying for grant money. So while the pool of funding has been relatively flat, the number of people applying for funding has increased dramatically. Increases in federal funding for science in 1998 (the first significant increase in over a decade) will only slightly decrease the level of competition for grants.

In the heyday of the Golden Age, federal agencies routinely funded more than 50 percent of the proposals submitted. In contrast, a proposal that I submitted to the Environmental Protection Agency in 1996 was one of 600 submitted, and only eleven were approved for funding. This acceptance rate was low even by 1996 standards. Typical acceptance rates were about 10 percent.

The decline in acceptance rates has had some significant implications. First, professors hired today face the same expectations to obtain grant money as when I came in 1990. The competition for grants is too great for this to be realistic. Thus, one can expect that fewer incoming faculty will receive tenure. For economic reasons, universities will likely feel compelled to keep their standards for grant success high. They will limit granting tenure to the occasional young faculty member who is a "rainmaker" and manages to overcome the low odds of success.

Second, the types of proposals that are currently being funded are conservative in scope. In the Golden Age, there was some latitude for funding "blue sky" research projects. Such projects had a potentially huge payoff, but also had a high potential for failure. Now that money is tight, there is political pressure to show positive results with every dollar spent. The end of the Golden Age of federal funding has created financial disincentives for working on cutting-edge, high-risk problems.

Third, the lower rates of success mean that faculty members have to write more proposals in a given year to pay for summer salary, research, postdocs and the occasional graduate student. I started to do this myself in 1996, writing six proposals in one year in an effort to overcome the declining odds of receiving funding.

If other professors are responding similarly, then even without faculty growth, the number of proposals submitted will continue to increase significantly. This also means that approval percentages will keep dropping as the number of proposals written by each scientist in a given year increases. Eventually the proliferation of proposals is bound to plateau because universities are not growing significantly and an individual professor is capable of writing only a finite number of proposals. Writing many proposals and knowing that most will not be successful is in some ways a valuable, humbling experience. Personally, however, I can think of more efficient ways to be humbled.

Even with a failed proposal, all is not wasted. For example, the preliminary work I did for my first NSF proposal was thorough enough to stand on its own as a research article. So after I sent out that proposal, I took the first half that discussed preliminary results, turned it into a research paper, and published it in a hydrology journal. I've learned to do this with other proposals, both those that have been successful and those that have been unsuccessful. I imagine some other scientists do this as well. This approach to grant writing allows you to kill two birds with one stone. You can publish and try to get grant money at the same time.

I have also learned to avoid missing proposal deadlines. Early on, I had to drive one more proposal up to Washington, D.C. But eventually I developed the ability to use the mail just like everyone else. It helped that I stopped using the departmental copy machine. It also didn't hurt that two years later Federal Express extended their shipping deadline to 8:30 P.M.

8

Why Research?

It was cold, windy, and snowing. I was in Yellowstone National Park during the spring of 1996 and I was working outdoors. Fortunately, I have always liked cold weather. I sat next to my computer equipment and my temperature probes alternately connecting wires and entering a computer program on a waterproof keyboard. Because my hands had been frostbitten several times while playing football in middle school and high school, my fingers swelled when I took off my gloves and worked on the electronics.

Five yards to my right, Daisy Geyser, a prominent feature of Yellowstone, erupted every ninety minutes or so. If the wind was right, I could feel some of its spray. Five yards to my left, a buffalo alternately munched on the sparse vegetation and lay down on the warm bare ground. I didn't like being this close to the buffalo. The previous night my coworker, an old friend, had read aloud from an odd book that consisted solely of gory accounts of people who had died in Yellowstone. The detailed stories of violent deaths from buffalo encounters were etched in my memory. Every minute or so I looked up to see whether this particular animal was showing any signs of interest in me. But his curiosity seemed only casual, and he kept his distance.

The park was closing in a week (it was March and I was using spring break to do research) and it was a good time to set up instrumentation to monitor geyser activity. For a solid month there would be no tourists or employees around and the National Park Service had given us permission to monitor some of the geysers around Old Faithful.

This project was not a typical one for me. Usually, my graduate students work with me and the projects are designed so that we can count on a publishable result. It is much easier to get grants funded when the proposed work is well defined and conservative in nature. Also, like most professors, I don't think that graduate students should work on risky projects and this work in Yellowstone was inherently risky in terms of its science.

We were trying to see whether changes in geyser behavior could be used to predict earthquakes in the region. The theory behind this work was that areas with hot springs and geysers would be especially sensitive to the small squeezing and stretching motions of the Earth that might take place prior to major earthquakes. It was admittedly a far-fetched theory, but it was one that many Chinese geologists and geophysicists believed and had used with arguably modest success in predicting earthquakes. I was working with a geophysicist from the Carnegie Institution of Washington and an old friend from the U.S. Geological Survey on this project. Each of us thought that the Chinese approach, although reviled by most scientists in the United States, just might work. At any rate, we thought it was worth a little of our time to pursue.

The first thing we needed to do was get some preliminary data on how frequently some of these geysers erupted and what, other than earthquakes, might cause changes in eruption patterns. That was why I was installing my computers and temperature probes. After a few hours of adjusting my equipment, I got up and prepared to move on to the next geyser. I looked at the buffalo one last time and hoped that the next site would be buffalo free.

I know that much of the public (including the undergraduate population) doesn't think that there is significant value in university research. The common view is that most research involves accumulating knowledge of little use to anyone. For example, I can well imagine someone reading the above passage and thinking that my research is probably futile (we will probably never be able to predict earthquakes, especially with geysers) and I shouldn't be wasting taxpayers' money on trips to Yellowstone. Obviously, my view of the project's value is different. And while it is true that it is a wonderful benefit that I get to go to Yellowstone (and I view myself a very lucky man to be able to work there), where else in America can you find geysers? If penance has any value, I should also note that I've spent far more time making measurements in the 100-degree heat of the farms of the Great Valley of California than I have at Yellowstone.

My little project in Yellowstone is only a small part of what I do, and it is a virtually microscopic speck on the university research scene. In 1997, according to the American Association for the Advancement of Science, the government was giving around thirteen billion dollars a year (about 1 percent of the total annual budget and 17 percent of all federal research money) to universities for research. Funding continues at about that level today. About seven billion dollars a year of that money goes toward research with health applications. A little over one billion dollars a year goes toward research with military applications. About two billion a year is spent on directed projects in areas as diverse as highway design to understanding the impact of petroleum spills on aquatic habitat.

So at face value, about 75 percent of university-based research money is spent on engineering and biological, physical, and social science research projects that are designed in some way to improve our country. When I hear the public, students, or congressmen deriding the worth of university-based research, I think that universities must be doing a poor job of communicating the worth of that research to our nation. The lion's share of university research provides manifest benefits to society.

Let's look at the remaining 25 percent of grant funding for universities. This money is slotted for what is essentially pure curiosity-based research. A tiny amount (several tens of millions of dollars per year and falling) goes to the humanities. The rest goes to the sciences (including the social sciences). It should be noted that even curiosity-based research sometimes pays economic benefits. For example, in the 1960s, geophysicists were at work understanding plate tectonics, one of the most profound "curiosity-based" discoveries of the twentieth century. This work, which showed that the surface of the earth is composed of distinct plates that move and interact with each other, had no application at the time. Since then, knowledge gained from plate tectonics has allowed us to find new sources of petroleum, natural gas, and metals.

Certainly, not all curiosity-based research pays dividends. It is probably true that we "waste" a significant share of this three billion on projects where no direct economic benefits can possibly be expected. Can't we afford to spend some money on research that will almost certainly be solely of intellectual value? Gaining understanding of our world for knowledge's sake has a palpable benefit. Obviously, we shouldn't go into deep debt paying for research like this and we don't.

We spend 0.2 percent of the nation's budget annually on curiosity-based research projects in the sciences and humanities. However, if the public and Congress think that this money is not well spent and wish to eliminate it, I can live and so can the rest of the university community. This money is awarded strictly out of the generosity of our government. We don't possess an inherent right to it.

There is much that is right with university-based research. It is a vital part of universities that can benefit the nation and enhance education. It's just that in the Golden Age, the government gave universities too large a share of the research pie. As a result, they lost sight of their teaching mission, and efforts in undergraduate education became (and still are) lackluster. We simply have too much of a good thing.

Like most of my colleagues, I love research. In fact, my prime motivation for becoming a professor was not to teach, but to be in an environment where I had the liberty to pursue a wide range of research projects. When I came to Duke, I was a professor ideally suited to the university model of the Golden Age. In most ways, I still am.

Research allows me to pursue my intellectual passions. Some of my research interests are motivated by societal needs, but for the most part it is simply a pleasant coincidence that much of what I do has real world applications. I have always been fascinated by how things work, and since my father was an avid fisherman, I spent a considerable amount of my youth outdoors on the water. A typical summer weekend involved a day trip to one of the many rivers and lakes around Milwaukee, Wisconsin. Our annual summer vacation was spent in northern Wisconsin fishing for walleye, northern pike and the near mythical muskellunge. Given that I'm not fond of biology, I never had any interest to study fish (I still like to catch them). Instead, I more or less chose to study the water in which they lived.

I like almost every part of the research process. I enjoy going out and collecting the data, whether it be in Yellowstone or Columbus, Mississippi (another research site of mine). My data from Yellowstone consisted primarily of computer-based records of water temperature as a function of time. The computers recorded water temperature every fifteen seconds at several geysers and when the water temperature next to the geyser rose, I knew that an eruption had occurred. So after a month of recording, I had several records of geyser behavior that I could analyze.

Ideally, it would have been nice to concentrate solely on this research for a few weeks straight and get all of the data analysis finished.

But my job requires a fair amount of juggling. There are other projects that need work, grant proposal deadlines to meet, graduate students to supervise, university committee, and departmental meetings to attend and, of course, students to teach. So even a relatively small research project like this is stretched out over time.

In August, I had to deal with a deadline that put this work on the front burner. I was invited to present the preliminary results of my Yellowstone work at the American Geophysical Union meeting in San Francisco, an annual meeting of geophysicists. I started to work on the data in earnest so I would have something worthwhile to say.

In my analysis of the data, I used a lot of mathematics. This too is something that I enjoy. As a child, I pleasantly dreamed about numbers. While I slept, they would fly like spaceships through the ether. With the Yellowstone data, mathematical analysis allowed me to find patterns in geyser behavior. I wanted to see whether there were any patterns that could be related to changes in such things as barometric pressure, precipitation, or tides. Such changes are known to squeeze and stretch the earth in small amounts. If geyser behavior is influenced by rocks being squeezed and stretched prior to earthquakes, then it should also likely be influenced by other factors that deform the earth. I also wanted to look for patterns that were unrelated to any observable cause and not predictable. Such patterns are called chaotic. If geyser patterns were predominately chaotic, then it is likely that influences due to earthquakes would be difficult to identify.

First, I performed a quick and dirty examination of the data. I was lucky. There were indications that some of the geysers were responding to the squeezing and stretching of the earth caused by changes in barometric pressure. Perhaps this effort at using geysers for earthquake prediction wouldn't be a wild-goose chase.

I worked on the data set for a few weeks until I felt I had some results suitable for a decent talk. Then I spent a couple of weeks more making graphs of my results suitable for presentation at the meeting. I would have only twelve minutes to speak (a standard speaking slot) and I had made more than twenty-five slides to accompany my talk. During the week before the conference, I spent some time refining my talk and reducing the number of slides so that my talk would not run overtime. (If you run over your time allotment, a red light in the room begins to flash, making you feel like you're about to get a speeding ticket. It's an embarrassing experience that is best avoided.) By the time I left for the meeting, I didn't have a written script, but I could have given my talk in my sleep.

* * *

Every field of study has at least one or two annual national conventions. For literature types it's the Modern Language Association annual meeting. For chemists, it's the American Chemical Society annual meeting. For those like me in hydrology or geophysics, it's the American Geophysical Union annual winter meeting in San Francisco (there is a spring meeting on the East Coast as well, but it is much more sparsely attended). It usually takes place early in December.

More than 6000 scientists, almost entirely from federal research labs and universities, attend this convention. Some come just to hear the talks and poster sessions. (For poster sessions, you don't give a speech, but make a poster that contains your research results. The poster is displayed in a large room for four hours, and scientists mill around the room looking at research results.) But participation at this convention is quite high. More than 4000 people in any given year either give talks or present posters. The number of attendees at scientific conferences like these grew (like science funding) exponentially until the late 1980s. Attendance has more modestly increased since that time.

Attendance at annual meetings like this one is extremely important for scientists and I attend the meeting almost every year. When we descend upon fashion-conscious San Francisco we stand out. Hydrologists and geophysicists are generally oblivious to trends in popular culture. I can easily separate many of the conference attendees from the native San Franciscans and the tourists when I walk down the street. The older scientists (almost entirely men) tend to wear sport coats or suits that come from another era and from a time when they were twenty pounds lighter. The young and middle age scientists are attracted to the rugged outdoor look of the 1970s and sport beards (for the men), flannel shirts, blue jeans, and boots. On rainy days, they wear parkas that look like they were ordered from a L. L. Bean catalogue.

The meeting has its social aspects at night, but during the day it's almost all business. Meetings like this provide a forum to present the latest advances in the field. While in theory it is possible to keep up with one's field of study by reading research articles, there is not nearly enough time in the day to keep up with all of the research published in journals. Attending a meeting is like reading "Cliff Notes" of other people's research. To the outsider, it must be a very boring sight. But for me and the other scientists, it can be the scientific highlight of the year.

In the late 1980s and early 1990s, the mood at this meeting was sour. Grant money was tightening, jobs were difficult to find, and the conversation in the halls among the attendees was full of complaints. But by the mid-1990s, the bitching and moaning had died down significantly. Neither the money or the job situation had changed, but many of us had simply become accustomed to it. The conversation had turned mostly away from complaints and back to science.

Before the meeting where I presented my preliminary work on Yellowstone geysers, I pored through the program of talks and posters. I circled the four to six talks per day that I wanted to attend. I would spend the rest of the time wandering in the poster room.

In my choice of talks to attend, I pay as much attention to the author as the title of the talk. I greatly respect the work of a few of my colleagues and would pay an admission fee to hear them give a talk, even one as brief as twelve minutes. They are so insightful in the presentation of their work that they are almost gurus. For me, listening to an excellent scientific talk is like listening to an inspired music concert or reading a great book. It can put a long-lasting grin on my face and its words will stay in my mind for months and years afterward. Usually, there is simply polite applause after a talk, but on rare occasions speakers are given standing ovations.

While I may never receive a standing ovation or have flowers thrown my way after a talk, I aspire to impress people with the quality of my work. The day of my Yellowstone talk, I was predictably excited. I walked into the room to hear the talk scheduled before mine. My old friend and Yellowstone collaborator from the U.S. Geological Survey was presenting research that was separate from our Yellowstone work. I looked at his sport coat and it looked vaguely familiar to me. Then I remembered why. When we were graduate students together, we were the exact same size. Between us, we owned one pair of wool pants and one sport coat. For meetings like this, the pants and coat would shuttle back and forth between us. The sport coat he was wearing that day looked like the one we used to share. It would have been a stretch for him to fit into the old coat, but he had not added too many pounds over the years. For me, it would have been impossible.

I walked up to give my talk. Sometimes, I get nervous giving talks at meetings like these and have difficulty getting the words out. I find the twelve-minute format very difficult, and prefer to give talks that have a more leisurely pace. But this wasn't one of those times. Probably, watching my old friend had eased me. I felt in command of my material and audience, and finished with a minute to spare (no flashing red light for me this time). I answered a couple of questions from the audience and left the room. That afternoon in the convention hall, a scientist whom I admire came up to me. "I saw your talk," he said. "It was refined and elegant work." I smiled. I hardly ever get compliments like this. It made my day, my week, and probably my month.

After the conference, I could have written a research paper based on my Yellowstone work. But I already had tenure, and I no longer felt the pressure to publish every bit of minutia that I had worked on. It is a legacy of the Golden Age that evaluation of research productivity is partly based on counting the number of papers someone has published. Hence there is pressure to slice up research results into thin pieces, with each piece forming a research article with one thin idea. This is essentially the equivalent of taking a piece of salami best suited for one sandwich and using it to make three or four sandwiches. The end result is not better science, but a further increase in the glut of papers already in scientific journals. In my quest for tenure, I did some slicing of my own and somewhat sadly, I encourage my own graduate students to do this as well. If they don't slice up their work into "least publishable units" (or LPUs as they are known in the scientific community) they will be at a distinct disadvantage with others in competing for jobs and tenure.

There is also pressure to publish results quickly. Before the Golden Age, one could (provided a competitor wasn't working on the same topic) put together definitive research results every few years. Today, such an approach is completely unfashionable in the scientific community. Instead, our papers have the feel of progress reports and are absent of polish. Even with tenure, I don't dare to wait until all my work is complete before I begin to publish papers.

Still, I decided that I needed more data before I could publish my Yellowstone results. The National Park Service generously agreed to let me monitor several geysers for a period of one year, under the condition that I find a way to make my monitoring systems invisible from public view. This request turned out to be relatively easy to accommodate and in 1997, I began to monitor in Yellowstone once again. An extremely helpful and energetic park ranger agreed to service my equipment once a month so I would only have to go out to Yellowstone twice that year (believe me, I would have enjoyed going out there every month, but I don't have the money or the time to do it).

In 1997, I spent time processing and analyzing this new data in preparation for writing a research paper. For me, writing a research paper is akin to telling a story. To be successful, it has to engage readers and have enough meat to hold their interest throughout. After a year of collecting data, I had enough material to write a research paper of which I could be proud. Eventually, this work may lead to a means to predict some earthquakes. Even if it doesn't, it is, to my mind, worthwhile. At the very least, this research gives us insight into how geysers work. In a small way and at a small cost, it increases our understanding of the world around us.

SECTION THREE

Campus Politics

9

Matchmaking

It was the spring of 1992 and I was listening to a job candidate give a one-hour talk on his research. He was short and slight of build. His voice wavered a bit as he spoke and it was clear that he was nervous. This was typical of job candidate talks and was almost expected. His talk, a staple of any job interview for a professorship, was undoubtedly the most important part of the interview process.

On paper, he had all of the right credentials. He had received his Ph.D. two years before from a prominent school. For the last two years, he had worked as a postdoc in the laboratory of a well-known professor from another prominent school. He had recommendations from professors at both universities attesting to his academic prowess. He had already published several articles in highly regarded journals based upon the results of his Ph.D. research. While on paper he was impressive, we had many candidates who looked just as impressive. It was a phenomenon of the post-Golden Age that we had far more candidates than were necessary to make a good choice. Being a department with a job opening after the Golden Age was somewhat like being one of the most eligible bachelors in the country and announcing that you were searching for a wife. It was a painful and somewhat arbitrary process to select the three candidates we would interview from the 150 applicants, twenty of whom had credentials that were equally outstanding.

It was four o'clock in the afternoon when the talk began and for the job candidate, it had undoubtedly been a tiring day. It started with breakfast at 7:30 A.M. with myself and other members of the search committee. (The committee was the group of faculty in charge of selecting candidates to interview. All of the faculty members of the department would eventually vote on which candidate would be offered the job.) At 8:30 he began a series of thirty-minute individual interviews with virtually every faculty member in the department as well as the dean. He had been shuttled from office to office, and by lunchtime he had seen about a dozen people. In the afternoon he spent an hour talking with graduate students in a group session. Then he met more faculty (from outside the department) in a series of individual interviews. He had arrived the previous night and would be flying home that evening. After this talk, the pressure would be off and he could relax a little bit.

There were thirty people in the room listening, almost all of whom were either professors or graduate students. Perhaps two or three people in the audience had read some of the job candidate's research articles and knew enough about the details of his work to make a critical assessment of its technical quality. Another five to ten (including myself) knew the area of this candidate's research well enough to make some evaluation of its significance to the field. The remainder of the audience didn't know much about the candidate's research or his research field. However, the impression that this job candidate made on this majority group would be very important. This group included most of the voting faculty and they would evaluate the candidate on his overall ability to give a talk that would interest and captivate those beyond his narrow field of specialization.

The criteria for judging interview talks have changed considerably over time. In the Golden Age, we would judge candidates almost exclusively on the intellectual caliber of the material presented. The rule was that the smartest person doing the smartest research won the job. Using this single criterion proved to be far from ideal. Sometimes we hired quiet, meticulous, and introverted researchers who had a difficult time teaching and selling their research to funding agencies. Sometimes we hired very smart people with very big personality problems.

By the time I came on the scene, universities had expanded their criteria to include a history of getting grants. We were more concerned than ever about money. Since about 1995, because university administrators have been increasingly pressuring departments to teach large numbers of students, we have been also looking for people who have at least the potential to teach well. (This has been the positive side of the influence of administrative pressure. The negative side is that, as discussed in an earlier chapter, we feel compelled to "teach easier" than ever before to attract large numbers.) Nowadays, we want the whole package. We want someone who is smart, doing interesting research, can communicate effectively to a broad audience, and has a winning personality. Our standards have risen with the quality of the applicant pool.

I watched and listened to the candidate present his material. I thought that he was doing a very good job. The research, while a little outside my own area of expertise, seemed interesting and important. Initially, he was somewhat unsure of himself, but as the talk progressed he gained confidence. His presentation was logical and the slides that he was using to show his data were clear and easy to read but not too glitzy. (Unlike the corporate world, we generally eschew Hollywood caliber presentation materials. They are seen as frivolous.) When the talk ended, he answered a few questions from some of those in the crowd who were knowledgeable in his field. He was articulate, and his answers seemed to satisfy the specialists in the audience.

My own opinion at the end of the day was that the job candidate had passed muster. His talk went well, and in my interview with him, he seemed like he had the ambition and drive to make it in today's academic world. We had two more candidates to interview, and one or both of them might outshine him, but he would be a good hire. However, when I talked to other professors the following day it was clear that he had not won over everyone. Maybe the specialists thought his talk was interesting, but he failed to excite many of those without much background in his field of research. I asked a friend what, exactly, was deficient in the presentation. "He did an OK job," he said. "But he didn't convince me that his work was important. He seemed too caught up in his own research to look at the big picture."

This was a disparaging evaluation. If the candidate didn't understand his place in the scheme of things, he would not compete well for grants. He might do well teaching graduate students in his specialty, but he might bore undergraduates to tears. I knew that my own area of research was too close to the candidate's expertise to allow me to judge his ability to impress a wide audience. I knocked him off the list of potential candidates. If the next two interviewees failed to excite the department, we could move down the list and bring in two more who on paper looked just as good. We could afford to be picky.

* * *

Quite a few readers may remember unpleasantly a time as a student when, like this author, they sat in a college class taught by a less than competent instructor and wondered, "How in the world was this professor hired?" They may suspect, since they have also been taught by wonderful instructors who left long-lasting positive impressions, that the process is entirely random. Perhaps the university president takes all the applications, throws them down a staircase and hires the person whose paperwork goes furthest down the stairs. Fortunately, it doesn't work that way. It's just that in the Golden Age research universities did not generally consider teaching ability when they hired faculty.

By circumstance, I have spent a fair amount of time hiring and trying to be hired. Over a ten-year period, I have served on several search committees in three different departments and have interviewed for seven academic jobs (three since I became a professor), four of which have been offered to me. Universities hire new faculty very carefully, and in the post-Golden Age, they hire better faculty than ever before.

In the nineteenth century, the process by which universities hired was often done on a local scale. For example, the first chemistry professor at Yale, Benjamin Silliman, was hired while he was a Yale law student. One day, the president of the university, a friend of Silliman's father, ran into Silliman on the Yale quadrangle. In the ensuing conversation, the president offered him a position as a professor with the contingency that he first go to the Medical College of Philadelphia and learn the field of chemistry.

Just before the Golden Age, a university like mine would hire new faculty almost strictly through the old boy network. A provost, dean, or department chair would call up someone they knew at another university and ask the standard question, "Do you have a good man for us?" (Women faculty members were rare at the time.) Sometime during the Golden Age, the process became much more open and the faculty were given a significant decision-making role. Deans and provosts decided in which departments new positions were allocated, but the faculty typically were in charge of the process from the start (writing the advertisement) almost to the finish (making a decision on a candidate to pass on to the administration for final approval).

Despite its openness, in some ways the Golden Age was a low point in faculty hiring. In the first place, it was a sellers market. There was a small pool of faculty available relative to the number of job openings. While the elite institutions probably had no trouble finding excellent faculty, lower level institutions had to settle for faculty who weren't always the best and the brightest. (You don't have to be a major intellect to get a Ph.D. The most important characteristic you usually need is dogged determination.) Also, the criteria for hiring, as noted above, were so skewed toward research prowess that we sometimes hired faculty members who were either poor instructors or had impossible personalities.

Nowadays it's a buyers market. Universities are no longer growing, yet we continue to pump out large numbers of Ph.D. graduates. (The number of new Ph.D. graduates increased 30 percent from 1986 to 1993.) When we place our advertisements in the magazines of professional organizations, we can expect to receive in excess of 100 applicants. As the Golden Age hires retire and are replaced, it is likely that the possibility of having a poor instructor will be greatly reduced. The market is currently so highly competitive that even mediocre colleges and universities are attracting excellent faculty who potentially can both teach and perform research.

I wasn't aware of just how incredibly competitive the job market had become until I served on my first search committee in 1992. In my own search for academic jobs in 1989, I had been fortunate. I sent out applications to only two institutions and was lucky enough to receive two job offers. Looking at the 150 applicants for the position available in 1992, I felt very fortunate to have a job. The pool of applicants was outstanding. The list of talented, ambitious, and productive prospective faculty members was very long. Perhaps half of the people on this list were, on paper, of better quality than faculty hired in the 1960s and 1970s at the university. But it was not possible or desirable to fire our senior faculty and create positions for these applicants. Out of the seventy or so excellent applicants, we had to select three for interviews.

We met as a search committee a few times looking at the applications and discussing the candidates. After the second meeting we voted to select the top twenty. After the third meeting we narrowed it down to the top five, the top three of whom we would fly in for interviews. The selection criteria were based on a number of factors. First, we restricted our search to those who had gotten off to a fast start and published a significant number of research articles in prominent scientific journals. If it had been a few years since they received their Ph.D., we wanted evidence that they had been successful in obtaining grants. The idea behind these criteria was that it was tough to get tenure and the person that we hired had to be on a trajectory to get tenure from the start.

Recommendations were also very important. In order to make the top twenty, the applicant had to have recommendations that consistently indicated outstanding achievement and promise. Especially helpful were recommendations from prominent professors or a recommendation that stated that the candidate had been the best Ph.D. student in a given department over the last several years.

In addition, the quality of the Ph.D. granting institution was considered significant. We were certainly biased toward those who graduated from the elite top ten research universities in our search. However, we also considered (and have often hired) applicants from institutions that are not considered among the elite in their overall reputation, but are considered to have world-class faculty in the area in which we wish to hire.

Finally, a problematic criterion that has come up on every faculty search in which I have been involved, but that we have usually managed to dodge, is the "buddy factor." It is inevitable that one of the search committee members will have a personal connection to one of the candidates. The candidate may be a friend, former student of a friend, or have a Ph.D. from the same university and department as the committee member. Usually, that committee member will strongly champion that candidate. When the candidate in question is not qualified for the job or is not competitive it can create unpleasantness in the search committee.

The worst case of the buddy factor that I have experienced was on a different search committee than the one in 1992. The son of a friend of the committee member had applied for a faculty position. Graham, the committee member, lobbied hard and gave a long speech on the "star potential" of this candidate. I tried my best not to roll my eyes.

"I don't get it, Graham," I said. "How can you say that someone who has no publications, no grants, and good, but not great, recommendations from an average school is going to be a star?"

Graham looked as if I had wounded him. "He has character," he said. "I don't think there is a single candidate who has as much character as he does. Above all we need someone with character."

"He's right," chimed in an old friend of Graham's who was also on the committee. "Character is very important. As a matter of fact, I read in a recent business publication that CEOs of large corporations look at character first and foremost when they hire."

"My daughter has character," I said. "She doesn't have a Ph.D. or any publications, but she is articulate and sharp as a tack. Maybe I should get her to apply for the job," I said. Fortunately, the other two members on the committee also thought that Graham's lobbying efforts were ridiculous. We were able to avoid the embarrassment of interviewing someone who, on paper, was well below the caliber of the best applicants in the pool.

On my first search committee, we also encountered the buddy factor. Someone on the search committee had a friend who was an applicant. But his lobbying for this candidate was only mildly beyond reason, and he had no old friend on the committee to back him. We batted away this effort with relative ease.

In 1992 our top three candidates were very diverse. We had the candidate described above whose interview went well, but not well enough. We had an older, much more experienced candidate who on paper looked slightly better than the other two (who also looked outstanding). But his talk was on a pedestrian topic, and was long on speculation and short on data. His interviews convinced the faculty that whatever enthusiasm for research he once possessed had disappeared. It was hard to reconcile such an outstanding academic record with the person who came to interview. An application, however, can only show a limited amount. It is a good thing that we don't shop for professors solely by mail.

By the time the final candidate came, I was somewhat nervous. We had already informally decided that neither of the first two candidates was acceptable. If this one did not work out, we would have to go further down the list. I felt we would still be able to find an excellent new faculty member, but the process would by necessity (given that we had not yet called anyone beyond the top three candidates) drag on for a couple of months longer. Fortunately, our last candidate proved more than up to the task. He was energetic, confident, and articulate. He had an amazing record of research for someone so young and recommendations that were stratospheric in their level of praise. When the department met to discuss candidates, the choice was clear.

Searching for new faculty is a process that I enjoy a great deal. It is always nice to get fresh blood into a department and it is especially nice because the quality of the post-Golden Age applicant pool is excellent. The searches have been, in the end, delightfully free of politics and chicanery. When we have made decisions, I have felt confident that we have truly found valuable and impressive additions to our staff. While it is a pleasant experience for me, I know that for those searching for a job it is painful. There are simply far too few academic jobs available for the number of potential professors. The market is poor for virtually everyone, from the humanities to the social sciences, to the natural sciences.

In the 1990s, the academic job market has evolved to something akin to the job market for actors. Instead of waiting on tables for their big break, science Ph.D.s take postdoctoral positions to fatten their resumes with publications and grants. We now have well over twenty thousand science and engineering postdoctoral gypsies, many working for low pay and some traveling from university to university every two or three years. Humanities and social science Ph.D.s try to find temporary low-paying jobs in such places as community colleges hoping that their ships will eventually come in.

There are some who argue that the current situation is best for universities and for research productivity. This argument is based on the principle of survival of the fittest. Because the market for jobs is so tight, potential candidates work harder than they would if it were a sellers market. There is truth to this argument. However, the market was competitive enough in the 1980s. Our level of competition is now so keen that it has become unhealthy for all. It has cast a pall over many university departments. Graduate students spend a good deal of time during their studies wondering just why they are working for a Ph.D. when the academic job market is so poor.

The problem of too much competition has recently been exacerbated by the reluctance of faculty hired in the Golden Age to retire. Due to a federal law enacted in 1994, universities had to eliminate mandatory retirement for faculty. I happen to like having older faculty around. They give my university a strong sense of continuity and its historical place in the world of academia. However, it would be better for universities if many retired voluntarily. (I know that I may feel differently as I approach sixty.) These retirements would create a little room for the many talented individuals nearly begging for the opportunity to become professors. Besides, most professors hired during the Golden Age have benefited handsomely from pension funds tied to a stock market that has boomed for almost all of the 1980s and 1990s. They can easily afford to retire.

The awareness of our troubled academic job market for recent Ph.D.s is widespread. In 1995, the National Academy of Sciences published a report from a panel commissioned to examine the state of U.S. graduate programs in science and engineering. Surprisingly, their findings indicated that we weren't overproducing Ph.D.s. Graduates in science and engineering were finding jobs, but they weren't finding the professorships that motivated them to go to graduate school. This report, however, does not paint the whole picture. The quality of and pay levels for many of the jobs occupied by recent science and engineering Ph.D.s is not good. In the humanities and social sciences, however, the situation is even more bleak. The auxiliary market for these Ph.D. graduates is very limited.

Yet some relief may be in sight. In the mid-1990s the number of students applying to graduate school began to drop. Perhaps students are choosing alternative career paths because they know about the lack of academic jobs. Or more probably, they are simply being siphoned into the job market because of our booming economy. Whatever the reason, if this trend continues, we many end up with a market that is competitive, but has a healthier balance between academic job seekers and job availability.

In the Golden Age, job availability was so high that it was often easy for a faculty member to move from one institution to another. This gave professors a great deal of leverage with administrators. They could go out shopping for another job and announce to their administration that they would leave unless certain demands were met. Given the relative difficulty of finding good professors in the Golden Age, administrators would often cave in to such demands. Getting job offers became a standard ploy to get a raise in salary, money for a technician, better laboratory space, better office space, and even better parking space. At Duke, occasionally a basketball-crazed faculty member with a job offer elsewhere would demand season tickets (which have been extremely hard to obtain since the 1980s).

The relative availability of jobs also provided political leverage for faculty. Before the Golden Age, the role of faculty in university decision making was usually marginal. It is not a coincidence that during the age of faculty mobility, university administrators almost universally yielded control to faculty on such issues as job searches and the undergraduate curriculum. With the end of the Golden Age, the ability of professors to find job offers that yield lucrative counteroffers from their host institutions has diminished markedly. Predictably, the pendulum of university power is swinging back toward the administration.

But obtaining a job offer from another institution is not just about money and power. Mobility has allowed for better matchmaking between professors and universities. Like corporations, every university has its own personality. They all have certain strengths and weaknesses in research and tend to attract undergraduate students that follow a certain profile. Professors who may be a poor fit in one university may find that their approach to research and teaching is well suited to another university. For example, one of my engineering professor friends was miserable at an Ivy League university because of its perfunctory interest and investment in engineering. He has since moved to a university with a strong engineering school and is quite happy.

When I came to Duke, I did not really know how long I would stay. I viewed my employment as a kind of trial marriage. If it worked out, that was great. If it didn't, I felt that I could probably find a job somewhere else. While most new faculty members in the post-Golden Age were lucky to find their one job, I was fortunate to be in a field of study that still afforded some mobility. Hydrology was an emerging field and many universities had historically not hired hydrologists. There was a minor nationwide trend of adding hydrology faculty at universities.

During the first couple of years, I did not have a strong desire to move. My university still thought it had a lot of money. It was trying to elevate its stature in the physical sciences and engineering, and as a physical scientist, I felt part of something that was rapidly improving. But eventually, the end of the Golden Age caught up with the university. It began to scale back its efforts to grow in the physical sciences. I was being treated very well, but it was clear that I was not going to be part of a growing and blossoming enterprise. So I began to take advantage of my mobility and shopped around. I wasn't trying to wrangle special benefits from my university's administration. I was trying to find a better match.

Mostly, I applied only if invited to apply. If someone with a job opening phoned, that meant that they knew who I was and liked what I did. It was possible to get a job by submitting an application without an invitation (in fact, all of the job offers on the search committees in which I took part have gone to uninvited applicants), but the odds were not as good. Since I already had my job and was being treated well, I was a careful shopper.

Visiting the college campuses during my interviews, I asked a lot of questions and tried to gauge just how pleasant and reasonable my colleagues would be. Each visit to another institution would more or less affirm that the match that I had already made, whatever its problems, looked pretty good in comparison. For example, I went to visit a university in the western United States. My wife and I love the West, and I thought that the potential for this match was good. However, during my interview, it was clear that there was little affinity, either scientifically or socially, between myself and those in the department. The air of political correctness hung heavy. When I went to dinner with one of the search committee members, he spent two hours telling me about his family problems. I sympathized, but I barely knew him, and his confessions made me feel extremely awkward. This university was good, but it was no place for me.

After I received tenure, however, I received a job offer that I thought I could not refuse. While my own university was still treating me very well, the other one was clearly going to treat me better. Its existing strength in the physical sciences meant that I didn't have to try and build, as I had done at Duke, a research program from scratch. They offered me a tenured position, an attractive salary and laboratory space, and plenty of money to purchase field and laboratory equipment. The faculty seemed pleasant and eager to have me come on board. I was all ready to sign on the dotted line.

But when I went to visit a second time with my wife, my dream match started to unravel. My wife noted that the town was small and the weather was cold. It was such a small town that our entire lives would revolve around the university. There would be little for her to do. As for me, she predicted that indeed this would be a great place for me to get work done. But it was likely that, in a setting like this, my tendencies to let work take over my life would get the better of me. "We'll be less than a mile from your office," she said. "You'll be there morning, noon, and night."

I wasn't happy to hear her tell me this. But she was right. My dream job would likely create a nightmare for our marriage. Whatever deficiencies were present at my university, I was in a setting that allowed me to balance work with family life. Besides, my job was probably better than about 95 percent of all university jobs that I could find. I'm in a prominent university with excellent students. It isn't a perfect match, but you can ruin your life by searching for perfection.

10

The End of the Golden Age

Every year in the fall, the president of my university gives a speech to the faculty. It is telling of modern university life that few faculty members attend these speeches. With the growth of the university, there has been a good deal of decentralization, and the importance to the faculty of the opinions or actions of a university president are usually not all that great. There is also a significant level of apathy concerning the activities of the administration amongst faculty. Besides, a synopsis of the speech is dutifully printed in the weekly university paper and the full text can be found on the Web. At Duke, the speech is given in the same 150-seat lecture hall where I taught my first undergraduate class.

By 1995, our president had already been at the university for two years. She had given many speeches on campus and her speaking style was well known. She tries to command attention by projecting dignity and forthrightness. On this day, she chose as her principal topic strategies for academic excellence for the university.

She noted that in the recent past the university had undergone growth in many areas. As a result, it had established itself as one of the top research universities in the country (hyperbole like this seems to be expected of presidents of all institutions and does not seem to negatively effect their credibility). Economic limitations, however, precluded the possibility of growth across all aspects of the university.

She outlined a plan for continued growth that she termed growth by concentration. We would select only a few strategic areas where infusion of support would likely yield the biggest benefit. Rather than try to be a university known for excellence in all fields, we would narrow our focus and identify a few core areas where we would have a worldwide academic reputation.

This speech did not go over well with the faculty. Our previous president had overseen economic difficulties but had never told his faculty that there were limitations to what our university could achieve. Faculty did not want to hear about "growth by concentration." But university presidents and other administrators, whatever their shortcomings (and they have many shortcomings), know the nature of the balance sheet and that the money just isn't there anymore. In her speech to the faculty, our president wasn't expressing the opinion that the good times when every department could expect growth had come to an end. She was stating a fact. Perhaps the prior reticence of our presidents to talk about fiscal limitations came from an awareness of the potential impact of such statements on their employment. When, in 1991, Yale president Benno Schmidt pushed for budget cuts, the resultant protest was so great that he had to resign. By 1995, however, the end of the Golden Age was old news and even faculty knew, mild protests aside, that the world of academia was no longer the same. There were a series of events, chief among them being the end of the Cold War, that had led to the demise of the Golden Age.

At the end of the 1970s, the American research university was a confident, financially healthy institution that by and large was held in high esteem. Federal funding of university-based research had stopped growing exponentially in the late 1960s, but support for educational programs was still strong. In response to the end of exponential growth in research funding, universities learned to stretch their incoming dollars by steadily increasing overhead rates (a tax applied to grants to cover the costs of running the university). Growth in tuition at private research universities, the other half of the revenue coin, continued at exponential levels and, as a result, so did the universities' overall rapid growth. For public universities, increasing numbers of students enhanced revenue. Only a true pessimist could have forecast the Job-like misfortunes that would befall the American research university in the 1980s. Some of the maladies of that decade that impacted universities were external in origin and others were (unlike Job's) self-afflicted.

Most of the early problems were not financial, but political in nature. The difficulties started with the rise of the conservative movement and the election of Ronald Reagan in 1980. Almost overnight, liberal political philosophy was only barely viable with the American electorate. True leftist political philosophy had been completely buried. For American colleges and universities, this transformation of the electorate created a severe public relations and credibility problem.

Universities (not including military academies, engineering schools, and those with a strong religious emphasis) have been unabashedly liberal since the Vietnam era. Natural science and engineering departments have probably always been more or less politically neutral if not politically agnostic. But those in the humanities and social sciences are largely dominated by liberal political thought and give universities a significant political slant. The liberal nature of universities remained unchanged in the 1980s and left them out of touch with the political leanings of mainstream America, including mainstream college students. If anything, universities hardened themselves politically in response to this external change. The eighties saw the rise of political correctness at universities where efforts at eliminating racial, sexual, and ethnic bias on campus sometimes became twisted into an intolerance of any political expression that was not from the left. The media, comedians, and politicians had a field day with universities' anachronistic political bias. While some of the criticisms were overblown, they contained kernels of truth.

If universities' problems in the 1980s were confined to being the butt of jokes in movies and television shows and condemnation by conservative politicians and journalists, all would have been manageable. Since the Vietnam War and Watergate, Americans have developed a love for criticism and ridicule of once-hallowed institutions, and it is not a tragedy that universities have been added to that list. At the end of the 1980s, however, universities were faced with more than just vulnerability to criticism of their political bias. They also had to confront criticism of how they used their federal funds.

The issue of use of federal funds began right under my nose while I was at Stanford University. You can trace a part of its origins to an electrical engineering departmental office and laboratory less than 200 yards from my own student office. At the time, I was busily trying to finish my Ph.D. dissertation and I already had a full-time job and a family. Like most scientists, I didn't pay much attention to campus politics, and I was so busy that I couldn't have paid attention even if I had wanted to. Unbeknownst to me, in that office and laboratory, one professor was not a happy man.

In 1987, John Madey was a research professor of electrical engineering and physics at Stanford University. He more than held his own in a university full of world-class scientists including many Nobel prize winners. His main claim to fame was his invention of the free electron laser, a technological breakthrough with the potential to transform aspects of military weaponry, manufacturing, and medical surgery.

Professor Madey's invention formed the backbone of President Reagan's proposed "Star Wars" defense initiative and he took full advantage of the growth in military research in the 1980s to obtain large amounts of research funding. By the mid-1980s, he became unhappy with the amount of his research money that went to university overhead. As noted above, universities had steadily increased the overhead rates associated with grants in the 1970s and 1980s in order to capture more revenue for general university operations.

Overhead rates vary from university to university. The trend is that the more celebrated the university, the higher the overhead rate. I guess that this trend reflects a notion that you have to pay the university a tax commensurate with its status. By the mid-1980s, overhead rates had reached levels of about 75 percent at top-notch universities such as Stanford. This meant that every dollar spent by a professor on research (with some significant exclusions) had a tax of seventy-five cents imposed by the university to be paid by the funding agency. Overhead rates were scheduled to rise even higher in the future.

To a university, overhead and tuition are its lifeblood. To a professor, however, overhead can be a major impediment to writing competitive research grants. "It feels like it's coming right out of the researcher's hide," said Nobel prize-winning Stanford physicist Steven Chu to the *Chronicle of Higher Education* in 1991.

Let's look at a simple example. Suppose you are Professor X from Stanford University and you write a grant proposal to a federal agency to do some research that will cost $100,000, excluding new equipment purchases. The total amount of your proposal including overhead will be $175,000 plus equipment purchases (equipment purchases are not usually subject to overhead). At the same time, Professor Y from University of Illinois writes a similar grant proposal with the same cost. Yet his overhead rate is only 45 percent, so the total cost including overhead will be $30,000 cheaper. Given equally good proposals and professors of similar reputation, the funding agency would be best served financially by funding the proposal from University of Illinois. So Professor X has lost a grant on account of high overhead. Or the funding agency still might prefer Professor X to do the research because it looks better on paper for the funding agency to have work done at the more prestigious institution. However, the funding agency thinks that the premium associated with Stanford too high. They will call Professor X and offer to give him 80 percent of the amount of money that he requested. Hence, because of high overhead, Professor X must find a way to do the work with 20 percent less money than he originally budgeted.

Technically, government agencies state that overhead should only be used for the administrative costs associated with supporting re-search. Universities were partly able to increase their overhead rates and expenses allowable for overhead in the 1970s to 1980s through creative accounting and finding loopholes in governmental regulations. Universities maximized their overhead by obtaining written agree-ments with federal agencies called "memoranda of understanding" to allow them to place costs not truly associated with research under the overhead umbrella. Stanford's overhead rate was high in comparison to typical universities and their number of "memoranda of under-standing" with federal agencies was far greater than any other uni-versity. Its aggressive stand concerning overhead was known through-out academia and its chief financial officer was known as "the sharpest pencil in the West."

While universities try to increase overhead associated with grants, faculty members, like many taxpayers, often try to find ways to reduce the amount of overhead they pay. For example at my university, no overhead is charged to single equipment purchases that cost more than $500. So the trick is to gather up small pieces of equipment until their total exceeds that amount and claim that each piece is one part of a major contraption of some sort. Overhead, like any tax, is a necessary evil and it often creates friction between faculty and administration.

Professor Madey was not the only faculty member unhappy with Stanford's rate of overhead. In 1988, there were protests from many other faculty members on this subject and a faculty task force rec-ommended that Stanford should become a national leader in establish-ing reforms to curb overhead rate increases. John Madey was unique in that he was internally audited because he utilized loopholes to max-imize the amount of money that went to research and minimize the amount of money that went toward general university operating ex-penses. The chief bone of contention between Stanford and John Madey was his use of non-Stanford employees and services. By hiring employees and buying services on contract, instead of within the Stan-ford system, Madey was able to save over one million dollars per year in overhead. The Stanford administration was infuriated by this action and closed the loophole on all of his current grants and contracts. Since his current grant and contract money was fixed, Madey was now faced with the prospect of paying more overhead without having additional grant revenue. He thought that this change was unreasonable, but to Stanford, these new overhead charges were essential to the economic health of the university. In 1988, Madey decided to leave Stanford and search for an institution with lower overhead. As coincidence would have it, he decided to join my university, which built a multimillion-dollar lab to house his research.

Madey did not leave Stanford quietly. He wrote letters to the editors of a local newspaper and a Stanford publication stating the nature of his dispute with Stanford. His leaving was noted in local newspapers, and the national magazines *Science* and the *Chronicle of Higher Education*. The university was nonplussed by all of this. It was one of the premier research institutions in the country in the Golden Age of the American research university. It was a confident institution with confident leadership. Subsequent to his leaving there were further disputes on how much of his equipment purchased on government grants could be moved from Stanford to Duke.

During the disputes between Stanford and Madey, the Office of Naval Research—the government agency in charge of oversight of Stanford's federal contracts—hired Paul Biddle as an auditor. Biddle was sympathetic to Madey's plight. He was an unusual man in many respects, including the energy and zeal that he carried to the mundane task of auditing Stanford's grants. In examining Stanford's records, he found many of the expenses unjustifiable. At first, Stanford dutifully complied with Biddle's many requests for paperwork. As the requests for paperwork mounted, however, the university administration began to belittle the auditor and question his motivation and competence. In response to Biddle's clashes with Stanford, the Office of Naval Research placed him on probation for 120 days. Biddle contacted Madey about his analysis of Stanford's overhead practices. Madey, who had already moved to Duke, referred Biddle to the office of Congressman John Dingell, chair of the U.S. Congressional Subcommittee on Oversight and Investigations.

In September 1990, the press reported that Biddle's estimates of misuse of overhead at Stanford were in excess of $200 million. A more detailed audit instigated by the Subcommittee on Oversight and Investigations found misuse of overhead funds that amounted to over $150 million. Overhead was used to pay for the maintenance of a yacht. Overhead was used to pay for flowers delivered to the president's house and for a formal reception introducing the president's new wife to the Stanford community. The national media gleefully broadcast the news that Stanford was using funds designated for support of research to finance luxuries.

Other universities like MIT and Johns Hopkins, fearful that Stanford's bad press would have a widespread impact, quickly announced that they would reimburse the government for past inappropriate use of grant money. Had Stanford rapidly followed suit, their troubles would likely have subsided. Stanford, however, stood firm. Throughout late 1990 and early 1991, Stanford's administration flatly denied that there was any misuse of funds. There were some accounting errors, such as the cost associated with the yacht. Other costs associated with luxuries, however, were perfectly legal as were the university's many memoranda of understanding. They justified Stanford's use of overhead using legal language that few could or cared to understand. They even maintained that Stanford had under-billed the government.

As the spring of 1991 approached, Stanford slightly changed its tune. While many of the costs charged as overhead, particularly those associated with luxuries in the president's house, were legally allowed in Stanford's eyes, they would reimburse the government for these costs. This action, however, came too late for Stanford to reestablish goodwill with the federal government.

On March 13, 1991, Stanford University officials were publicly grilled and lambasted by both Republicans and Democrats during a seven-hour congressional hearing. In viewing videotapes of the hearing, it is obvious that there is no common ground between the congressmen and the Stanford officials. Stanford officials act as if they are being victimized. "There is no evidence brought out that we did nothing other than to follow government law," said Stanford's president. "I'm concerned about loose implications of criminality about people at Stanford."

To the congressmen such forceful language smacked of defiance and arrogance. "It really disappoints me to see a university president . . . stonewalling this committee like a common politician," said Representative John Bryant when Stanford's president refused to acknowledge that Stanford's financial practices were aggressive. The tension between the congressmen and the Stanford officials was palpable during the hearing. Stanford's representatives acted as if the hearing was a formal debate that they could use to argue their case.

Stanford's president, a longtime faculty member and former head of the Food and Drug Administration, lectured and used graphic displays to try and show the necessity of overhead for the nation's research. He noted that while overhead rates rose over the previous decade, the actual percentage of money that went to university administrative costs was largely flat. But as Representative Ron Wyden, a Stanford graduate, noted, "This meeting is not a debate about the details of overhead." The congressional hearing, in essence, was a public ceremony to formally rebuke universities across the country and Stanford in particular for their billing practices. During the hearing, Stanford's president tried to control his anger over accusations that he viewed as unjust. On the other side, the anger and disgust of the congressmen was unmuted. It was as Representative Wyden stated, "a very sad day for one of the world's great universities."

Ironically, Stanford was mostly financially vindicated in the end. Subsequent government audits found Stanford guilty of only minor infractions. Their billing practices, such as billing the government for a computer system bought for private fund-raising, looked inappropriate and were a public relations disaster, but they were almost all legal. Hundreds of millions of dollars were not being used illegally. However, it wasn't until three years later that almost all of the original audit results were reversed (Stanford was forced to return $1.2 million in addition to the $1.8 million they reimbursed the government in the winter of 1991). By that time, the president of Stanford had long since been forced to resign. As noted by Jerrold Footlick in his book *Truth and Consequences*, Stanford's legal and accounting costs associated with their overhead scandal totaled about $37 million.

In the intervening years, universities across the country were audited and forced to roll back their overhead rates. At Stanford the overhead rate was initially reduced by 30 percent and as a result, they temporarily suffered annual budget deficits of tens of millions of dollars. At other institutions the reduction in overhead was not so severe (at my university it was less than 10 percent), but the revenue loss associated with this rollback totaled millions of dollars per year for some major research universities. The nationwide audits also resulted in universities collectively refunding millions of dollars to the federal government. To make matters worse, at the same time that the overhead scandal spread, twenty-three of the elite private universities and colleges of the Northeast were being sued (privately and by the Department of Justice) for the price-fixing of tuition. The flip side of growth of research universities was that they were increasingly looking like just another large business sector with shady financial practices.

As a result of the overhead scandals, the era of universities increasing overhead rates over time to enhance operating revenue came to an end. With it also ended the memoranda of understanding that allowed universities blanket privileges to spend overhead on items not directly related to research. For example at my university (a very minor overhead offender) overhead had been used to partly subsidize the cost of classical music concerts on campus. Stanford similarly included student union costs under its overhead umbrella. After Duke was audited, I found myself paying a couple dollars more to hear music. Perhaps the logic behind using overhead for this purpose was that after a good concert, I might be inspired to do better research. The bottom line was that without being able to increase overhead costs, universities were forced to become more fiscally prudent. To avoid future embarrassment associated with their billing practices, they also had to dramatically increase their accounting staffs. True fiscal restraint, however, was just around the corner. It came with the end of the Cold War.

In the fall of 1990, about the same time I began my job, I received an odd e-mail message. Someone sent me an electronic advertisement for the sale of pieces of the Berlin Wall. Back then, the commercialization of the Internet was truly in its infancy. It was unusual to receive any advertisement, much less one selling pieces of rubble. Apparently for the price of seventy dollars, I could buy a "piece of history." The Berlin Wall had been knocked down the previous year and some enterprising person took advantage of the opportunity to financially benefit from the destruction of this Cold War symbol.

At the time, I didn't bother to think that the end of communism in eastern Europe might have a significant impact on university life in America. But that purported chunk of the Berlin Wall that I could have bought in 1990 with my Visa card was also, in a strange way, a symbol of the Golden Age of the American research university. The Cold War had compelled the federal government to engage in massive spending on scientific and engineering research, much of which was performed at universities.

The rationale for this spending was principally that the United States needed to maintain military superiority over the Soviet Union and be the undisputed leader in science and technological development. The Soviet Union responded in kind, and the stakes seemed to escalate every year. The results were record budget deficits in the United States and ultimately the complete economic collapse of the Soviet Union.

Since the decline in the fortunes of our former arch-rival, the United States has followed a much more cautious approach to federal spending on research, both military and nonmilitary in nature. As a result, universities have had to contend with long-term financial constraints for the first time since the end of World War II. Our nation's recent projected budget surpluses have eased the financial strain only slightly. The end of the Cold War meant that the era of growth in universities through federal funding had come to a close.

The impact of the end of the Golden Age has been particularly hard upon those universities with a large investment in the physical sciences and engineering. These areas were heavily supported by grant money from the military during the Cold War. Departments in the physical sciences and engineering with strong research emphasis had been major sources of overhead that was used to support other aspects of the university. But science and engineering departments require large amounts of space for laboratories, technicians to run equipment, and constant upgrading of laboratory facilities to stay modern. With the flattening of research funds, the cost of maintaining these departments has in many cases exceeded their revenue.

As a result, prominent universities have decided to stop spending the money necessary to be at the forefront of some expensive fields of research. For example, Yale's President Richard Levin wrote in 1996, "No university has the resources to be the best in the world in every area of study" (*Yale's Fourth Century*). Unlike the previous Yale president, Benno Schmidt, he did not press for budget cuts. Instead, he recommended a policy of "selective excellence" and pointed to fields in engineering and the physical sciences where Yale would no longer try to compete head-to-head with the best. Needless to say, the response to this shift in policy amongst faculty members at Yale in these fields was not positive. But the uproar was decidedly muted in comparison to the response caused by the policy changes recommended by Schmidt.

My own general area of study, hydrology, has virtually no military or direct industrial applications. Its funding has typically come from agencies concerned with environmental issues. At face value, one would think that neither the Cold War nor its end would influence funding of research in this area. But the Cold War produced a halo effect that provided escalating levels of funding for all of the sciences. And the end of the Cold War has produced a negative effect on all of the sciences. Also, it is not clear that funding in environmental science research would have continued to grow dramatically, even if the Cold War continued.

Not only has federal funding flattened but it has also changed focus. In the past, the government had applied much of its research funds to ambitious, large-scale projects. The idea behind this approach was that breakthroughs are almost always the result of work with high risk.

But with the tightening of research funds has come an expectation of rapid positive results from researchers. Funding agencies, with the exception of those involved in funding medical research on major diseases, have generally chosen to focus their attention on small, safe, and inherently less ambitious projects.

Probably the most visible example of the impact of this change can be found at Princeton University. For more than forty years, it housed the premier facility for research in nuclear fusion in the country. At its peak, the fusion laboratory employed more than 500 scientists and engineers and had an annual research budget of more than $70 million. Progress in developing fusion technology proved to be exceedingly slow and in 1997, the laboratory became a victim of the end of the Golden Age. The facility was recommended for closure by a congressional panel.

Mostly, however, the economic changes at universities have not been dramatic. The Golden Age has been followed by a long-term gradual erosion of resources. Departments can no longer count on replacements for faculty who retire, leave, or are denied tenure. The size of support staff in departments (secretaries and technicians) has been reduced slightly. Salaries still rise, but at a level hovering around the rate of inflation.

What is most significant is the change in mood of faculty and administrators on campuses. The end of the Golden Age has caused severe erosion of collegiality and trust amongst faculty and between faculty and administration. Administrators at universities complain publicly that faculty salaries are too large a portion of the overall budget. The faculty complain that there are too many administrators and that administrators are overpaid. Universities have become a less pleasant place in which to work, although the same can be said of much of corporate America.

Despite our economic difficulties, we have not, however, given up on the idea of growth. It's just that our models of growth have changed to accommodate the change in economic climate. For example, in her 1997 annual speech to the faculty, Duke's president discussed what we could learn from a corporate style of management. The "growth by concentration" motto of her 1995 speech was transformed into one that emphasized growth by substitution. Whatever growth we achieved in one area would be counterbalanced by pruning and shrinking in other academic areas.

The Golden Age represented an incredible run of prosperity for universities. There was a societal purpose to much of that growth as well.

First, there was the need to grow to provide higher education to the children of the baby boom. There were more children of college age than ever before, and a higher percentage of them wanted to go to college. Without growth, we would have been forced to aggressively limit access to higher education precisely at a time when the public demanded greater access.

The growth in research and graduate programs also had a societal purpose. It is common for those outside of the university to deride universities' emphasis on research and graduate education. But university research provides a critical training ground for those who go on to take jobs in a broad array of fields that use and develop technology. During the Golden Age, America needed a growing pool of people with expert training in engineering and science. Our universities were able, through their growth in graduate programs, to provide government and industry with an ample supply of talent. That talent, both American born and the foreigners who decided to stay in this county, allowed America to maintain technological superiority in many key areas, both military and commercial. Silicon chips, weaponry, pharmaceuticals, tele-electronics, medical diagnostic equipment, and other products are dependent upon a large workforce with research experience and graduate education. America benefits economically from having the best educational system in the world for graduate study in the sciences and engineering.

But how big do our research efforts have to be? At some point additional federal money for research provides diminishing returns to society. It is likely that America was receiving diminishing returns from its investment in universities well before the end of the Golden Age. Certainly, there were signs of excess. Congressmen were taking advantage of the largess to direct research funds toward universities in their districts with little evaluation of the quality of the proposed research. Liberal arts colleges and satellite state university campuses that did not have the resources or the type of faculty to perform excellent research were chasing after state and federal research dollars in earnest. Schools added Ph.D. and M.S. programs in efforts to increase their prestige without much thought as to whether there were enough talented students to fill these programs or demand for these students once they graduated.

By the 1980s, university research programs had grown to such a degree that there was significant bloat. For example, take geology, the field in which I studied as an undergraduate. At the end of the Golden Age, there were over 300 geology graduate programs in the United States producing 1600 M.S. students and 550 Ph.D.s per year. I happen to love geology, but there was not a societal need for this many geologists. There were not enough talented people wishing to go into this field to produce this many worthwhile M.S. and Ph.D. theses. The same can be said for almost all fields in the natural sciences and social sciences, and probably all fields in the humanities. While the Golden Age resulted in an explosion of excellent research and graduate student training, it was also accompanied by an explosion of mediocre and poor quality research and graduate student training.

It should also be noted that the Golden Age of the American research university was not the Golden Age of undergraduate teaching. Student populations grew dramatically until the middle of the 1970s (they are still growing, albeit much more slowly) and this growth helped fuel the revenue stream of universities. In the Golden Age, however, students and undergraduate instruction often received short shrift. Faculty became overly focused on research funding and publishing research results. University administrators encouraged this behavior by linking tenure and promotions almost entirely to research productivity. Research universities also erred by hiring a significant number of faculty who were outstanding researchers but had little affinity for teaching. High quality instruction was dependent upon faculty who felt an obligation and desire to teach undergraduates regardless of a lack of incentive and how much time teaching distracted from research. Fortunately, there were (and still are) a great number of these faculty, but there were also many faculty who did a poor job of instruction.

The Golden Age allowed America to become an intellectual leader in the social sciences, natural sciences, and engineering. But it created a university-based research program that is in many ways too large for its own good. The Golden Age is gone, and it is just as well. There is no longer a need for further growth. But there is a pressing need to fix the problems associated with the rapid growth that occurred over a period of forty years.

11

Shaking the Tree

I was visiting a large state university and sitting in an on-campus eatery with one of their chief fund-raisers. His specialty was corporate fund-raising. He was extremely likeable and unpretentious. He was the kind of person whose company you'd enjoy at the local watering hole for an after-work drink.

While we waited for the waitress, he told me a story. He had a friend who worked for a local charity and once went with her on a fund-raising visit to a wealthy executive. While his friend and the executive exchanged information, he sat silently. Finally, after about twenty minutes, the executive stopped his friend in midsentence. "This is all well and good," the executive said, "but I do have a question. This man you've brought with you. He hasn't said a word. What does he have to do with the charity?"

"I don't work for the charity," he said. "I do, however, think it's a worthy cause."

"So why, exactly, are you here?"

"That's a good question and it deserves a good answer. It's simple. I'm here to ask for the money." The executive chuckled and ended up writing a nice fat check for the charity. Asking for money is a talent. You have to have a unique combination of personality and nerve that makes people feel good about giving money away.

The waitress came to our table and asked for our drinks. I requested a Coke, but she said they didn't have any. "Would you like a Pepsi?" she asked.

"Sure," I said. I couldn't tell the difference between the two anyway. Nowadays, I prefer plain water.

"I'll have a Pepsi, too," the fund-raiser said and the waitress went to get our drinks.

"That was funny, you ordering Coke here," he said, smiling. I asked him why. "This campus is 100 percent Pepsi," he said. "You can't find Coke anywhere. We signed a deal with Pepsi for exclusive rights to provide soft drinks." He was proud of this deal and had played a large role in its development and consummation. For a multimillion dollar donation from Pepsi, his university had turned itself into a Coke-free zone for ten years. There were no Coke vending machines. There was no Coke in any of the dormitory, faculty, or staff eateries. The 100,000 plus football fans who came to the home games drank Pepsi unless they brought their own from home. "It's a good deal for both parties," he said. I was inclined to agree. Every year, Pepsi was given a good shot at making 80,000 students (including satellite campuses) Pepsi drinkers for life. In exchange, the university received a nice chunk of money to spend on a variety of items including scholarships and education. Of course, the university was eliminating the possibility of some students and faculty finding the soft drink of their choice. But not providing someone with their first choice of overpriced carbonated sugar water cannot be considered a significant loss of freedom.

Later that same year, back at my university, I was walking toward my office. Like most universities, there are posters stapled on billboards all over campus advertising a variety of items. Included in that week's roster of posters was an eye-catching composition of computer graphics. The Coca-Cola logo was in large print at the bottom. The name Coca-Cola or the corporate logo appeared on the poster four times (and three more times on the flip side for good measure). At first glance you might have assumed that the poster was advertising some product associated with the soft drink manufacturer. But if you stopped to read the text, you would have found out that the poster was promoting the opening of the university's Center for Environmental Education. The center is part of the School of the Environment and is designed to help elementary schools, high schools, and corporations expand their environmental awareness and expertise. What does this have to do with Coca-Cola? The Coca-Cola Foundation donated one million dollars to help create the center.

Universities have an unhealthy tendency to be insular, and it's a good thing when they develop outreach programs such as this one. While the center sometimes confuses environmentalist advocacy with environmental education, the Coca-Cola Foundation has given money to a good cause. But Coca-Cola, like Pepsi, does not give money away without getting something in return, and what they receive in this case is quite beneficial.

In the first place, Coca-Cola gets a little advertising on campus. The advertising benefit is not a one-shot deal related to the opening of the center. Associated with the Center for Environmental Education is a biannual "Coca-Cola Seminar Series." As a result of the donation, twice a year the university will invite a prominent environmental leader to campus to give a lecture. And twice a year, Coca-Cola will have the opportunity to insert corporate advertising onto a poster ostensibly giving information on the time and date for the lecture.

This kind of benefit, however, is small in relation to the size of the gift. There must be something other than a little local advertising that Coca-Cola gains from this money. And there is. When the gift was announced to the press, the administrators of the School of the Environment praised the company for its "long-standing leadership in the environment." By its donation, Coca-Cola has been given an environmental blessing by a respected institution. It's a blessing that Coca-Cola (and Pepsi for that matter) can well use. In the 1970s and 1980s, the soft drink industry collectively spent millions of dollars across the country fighting against "bottle bill" legislation designed to reduce refuse associated with soft drinks. Through its sponsorship of this center and its annual lecture, Coca-Cola is able to partially atone for its past and future environmental misdeeds. It may also have established a nice toehold on my campus for future exclusive soft drink distribution rights. We may not have 70,000 students, but even a market of 10,000 student customers is probably worth chasing after.

The search for donations by universities goes well beyond trying to profit from the cola wars between Pepsi and Coca-Cola. With tuition and federal dollars proving inadequate to cover costs, donations likely represent the only means to maintain the grandeur of the Golden Age. Most elite universities have endowments in excess of two billion dollars, but even billions aren't enough. Since the end of the Golden Age, universities have dramatically increased their efforts to find nongovernmental sources of funding. Most every adult who graduated from college knows very well that universities in the 1990s are soliciting their alumni more heavily than ever before. The additional effort requires more labor and universities have significantly enlarged staffs associated with fund-raising for both private and corporate donations. For example, Duke's School of the Environment had a staff of two associated with public relations and fund-raising in 1990. In 1997, it had a staff of a half dozen and the costs for public relations and fund raising accounted for roughly 10 to 15 percent of the total budget, an amount typical of many private charities.

The increased role of fund-raising has created a need for image enhancement and control of the flow of information to the press. The change probably began at the end of the Golden Age. For example, in 1989, longtime Stanford News Service head, Bob Beyers, resigned because his policy of candor with the media was out of step with changes in public relations at the university. By the mid-1990s the influence of image enhancement and public relations at Stanford was strongly evident. Its 1995 Strategic Communications Plan suggested employees use words like "boundless, challenging, incomparable, western/pioneering, stunning, and vibrant" to describe their university to the outside world.

The effort to raise money has had an impact beyond the role and size of public relations and fund-raising staff. It has completely transformed the role of administrative leaders on campus. Presidents and deans can no longer concentrate their efforts on running the university. They are increasingly representatives of the university on the fund-raising circuit and travel far and wide seeking donations. In addition to administrative skills, they must now, like the fund-raiser noted above, know how to ask for money.

The modern fund-raising president is entirely unlike the president of yesteryear. Because the president's key role is as a fund-raiser, he or she must delegate responsibility extensively. Presidents are campus leaders largely in a figurative sense. Not only are they disengaged from the workings of the university, they have also dropped their traditional role as commentators on the political, moral, and ethical issues facing society. Nowadays, out of fear of offending potential patrons, they avoid discussing any issues not related directly to higher education.

The formerly standard model of the stuffy dean in the ill-fitting tweed sport coat is also becoming a thing of the past. The dean of today needs to possess an easy smile, good clothes, a firm handshake, and is required to have cocktail party smarts.

Because presidents and deans spend so much time fund-raising, they must add new administrative staff to do the day-to-day work they once performed themselves. For example, in 1990 the School of Arts and Sciences at my university had one Dean of the Faculty. By 1997, this post had been expanded to three positions partly to allow one of them to go out in search of donations.

The goal of much of these efforts is to find the funds to increase the university's endowment. These efforts have been highly successful. It is fortunate that at the same time that the government has become fiscally cautious in its funding of universities, both corporate profits and the stock market are booming. As a result, corporations and individuals can afford to be generous. Also, the endowment funds are partly invested in the stock market and real estate so the economic boom of the 1990s has caused the endowment to increase dramatically on its own. When I came to Duke in 1990, its endowment stood at $500 million. In 1997 alone, the university raised over $200 million in private and corporate donations, and its endowment crossed the $1 billion barrier.

Other universities have also been highly successful at raising money and increasing their endowments through investments during the 1990s. The size of our endowment is about in the middle of the twenty or so elite private schools in the country. Universities typically use 3 to 5 percent of the total value of their endowment to help pay for yearly expenses (increases in the endowment above this amount are reinvested). Hence my university, as a result of endowment increases, has about $20 million more to spend from its endowment in 1997 than it did in 1990. For comparison and to express a little envy, Harvard University has seen its total endowment increase by $7 billion over the same time period, allowing for about $250 million in extra spending per year. While our increase in endowment money has served to partly cushion us from flat funding of grants, Harvard is probably alone among universities in that its endowment should allow for virtually complete insulation from changes in federal funding. Other elite universities may never equal Harvard's level of comfort, but through fund-raising they are trying their hardest to at least approach it. Harvard, ever trying to stay on top, also continues to push for more donations.

The flip side of the aggressive fund raising and newly acquired wealth is that they are hard to reconcile with the magnitude of tuition charged and the continued efforts to raise tuition at rates that significantly exceed inflation. The size of the endowments of the elite private universities, ranging from one to thirteen billion dollars as of 1998, now rivals the worth of some large corporations. Harvard's endowment is large enough that if only one percent more was spent annually and allocated toward undergraduate instruction, Harvard could eliminate tuition altogether. Similarly other elite institutions could significantly reduce tuition if they took advantage of their recently acquired bounty and simply spent a slightly higher percentage of their endowments. Instead, they choose to reinvest their money for a rainy day. Given the daunting cost of attending college, it's very clear that it's already raining.

To be fair, there have been some efforts to make college affordable again. In 1998, several elite colleges and universities (including Stanford, Harvard, Princeton, MIT and Dartmouth) announced the creation of enhanced financial aid packages for both the poor and middle class. Also, recent fundraising efforts at many colleges and universities, John Hopkins being the most notable example, have been partially directed toward increasing the pool of money available for financial aid.

Ideally, gifts to universities are unrestricted. But most people and most corporations want a say in how their money is spent. Strings are commonly attached to donations and they pose two central problems. First, there is the question of who runs the university? If someone from outside the university gives money to enhance a particular program, they strongly influence the direction of a university and in essence, they become part owners. Increasingly, as more money becomes donated with restrictions, university faculty and administrators have less independence. To a great extent, the problem of lack of independence was already encountered in the Golden Age. Rather than private donations guiding the direction of universities, it was federal grants that drove decision making and led directly to the growth of the sciences and engineering. Universities' continued dependence on federal grants has made the federal government a nearly full partner in the running of research universities. Increased dependence on private donations will give both wealthy philanthropists and corporations a larger say in how universities are run. Trading independence for financial comfort is probably unavoidable.

The other key problem associated with private donations (both restricted and unrestricted) is that they can compromise the integrity of the university. Obviously, trading integrity for financial comfort is disastrous. Universities through their actions are supposed to serve as examples of moral and ethical behavior for their students and for their alumni. When a university enters into a financial agreement with anyone, it should do so in a way that does not expend its moral and ethical capital. However, as universities become increasingly dependent upon private donations, it is probably inevitable that many will cross the line of good judgment and have their integrity at least temporarily scarred. As a 1988 *New York Times* editorial noted, "Oscar Wilde could resist everything except temptation. University presidents, it seems, can resist everything except money." The overhead scandals of the 1990s likely will be a prelude to a future period of donation scandals.

Currently, ethical problems with private donations at universities are generally of minor consequence. Private universities often show preference to applicants who are wealthy or whose families have already given significant donations. Sometimes the pull of money is so strong that universities admit wealthy students even though they are not close to being competitive with the rest of the academic pool. By establishing such a policy, these universities have, in essence, institutionalized bribery. However, the percentage of students admitted via this route is very small, and similar to admitting academically marginal athletes, many universities find this level of corruption to be acceptable.

Problems with corporate donations have been relatively minor as well. For example, consider the Coca-Cola grant received by Duke. Nowadays a one million dollar donation to the university is relatively small potatoes. How much ethical capital is the university sacrificing for this one million dollars? First let's look at the bulletin board space we give once or twice a year to Coca-Cola. I may find this exchange of money for advertising tacky, but public acceptance of advertising in all facets of life is common. People even proudly buy and wear clothing with corporate logos. So the advertising issue is a probably a minor one.

Then there is the issue of the environmental legitimacy gained by Coca-Cola through its sponsorship of an environmental education center. How bad is this arrangement where we participate in corporate image cleansing in exchange for cash? It depends upon the nature of the company. Coca-Cola is not an environmental leader, but it has spent considerable sums of money to reduce the waste associated with its products. It's poor judgment to engage in such a financial transaction, but it is not the stuff of a major scandal. One would have to come up with a more egregious and ironic corporate sponsorship before a major protest would ensue, such as a Dow Corning Center for Cosmetic Surgery Research. As outlandish and improbable as such an example sounds, I note that the Exxon Foundation (can we say "Valdez") was heavily solicited to be the prime sponsor of a marine sciences education center at a major university. Exxon wisely balked at this opportunity to participate in ironic philanthropy.

Increasingly, universities expect vendors to make donations. The previously noted Pepsi donation at a major university in exchange for exclusive distributorship on campus is for all intents and purposes what is known in the business world as a kickback. But at least the arrangement is out in the open for all to see. More problematic are the many small vendors of universities that are not in the public eye. For example, if a catering company who does business with a university is solicited for a donation, it is probable that the company will assume, rightly or wrongly, that a donation will increase its chances of continuing to obtain steady business. Also problematic are donations from members of a university's board of trustees who are executive officers of potential major vendors to the university. If a CEO from a major phone company were to make a significant donation at a time when a university was soliciting bids for telecommunications services, the donation would be seen as a bribe. Likewise, if a university president served as a member of the board of a large corporation that received contracts from the university, there would be a direct conflict of interest. So far universities have been circumspect enough to avoid major fund-raising scandals associated with their vendors.

Corporate grants to support specific research projects also potentially compromise university integrity. Of course, often such grants pose no problems. For example, oil companies have in the past directly supported some of the research of my geologist colleagues. The professor may be interested in performing some scientifically valuable research on the geology of an area that contains oil-bearing rocks. The company thinks that the geologic research will help them in their evaluation of the suitability of the area for oil production. Because negative information on suitability will be just as valuable to the company as positive information no pressure is placed upon the professor to come up with a specific result in research. As long as the oil company does not restrict the publication of the professor's findings, such financial arrangements are beneficial to both parties.

But corporate grants to support research are not always free of potential conflict. This is particularly true for research in the environmental and health fields. For example, a statistical study of seventy research articles on a controversial class of drugs used to treat heart disease was published in 1998 in the *New England Journal of Medicine*. Ninety-six percent of the authors with supportive articles had financial links to the companies that manufacture these types of drugs. In comparison, only 37 percent of the authors of critical articles had financial links. Universities and faculty members, in their desire to keep the research funds flowing, have ignored this potential problem with corporate research funding.

There is also the problem that even without potential conflict of interest, corporate sponsored and directed research projects are often very mundane in nature. For example, I know of one professor at a university who obtained several million dollars from oil companies to perform routine geologic investigations. He was funded to perform the work because it occurred in countries wary of oil companies and oil company geologists could not obtain entry. In essence, he and his students were a geological consulting company within the university. The "research" they performed was of marginal scientific value.

A 1996 report published in the *New England Journal of Medicine* analyzed the tendency for corporate sponsored research to be mundane. The report found that scientists who received more than two-thirds of their research support from industry published fewer research papers than those who received less support and those articles that were published with industry support tended to have less scientific impact. I also have had the opportunity to perform work on uninteresting projects that met specific corporate objectives. I have declined to do this work. Routine work should be left to corporations and consulting firms.

The lion's share of private and corporate donations are not research related. At the same time that universities evolved into institutions that were primarily oriented toward research (or developed a dual primary role of research and instruction), the public continued to view universities as institutions whose primary mission is undergraduate and professional instruction. As a result, when both private individuals and corporations think of giving money to universities they strongly prefer to support teaching and activities related to undergraduate life rather than research. This creates a strong conflict between university fund-raisers and professors. Professors would like fund-raisers to concentrate their efforts on finding funds for graduate-student scholarships and equipment for research. But such funds are hard to obtain. It is easier to find donations directed toward areas such as scholarships for undergraduates and computing resources for teaching.

In 1994, I was directly impacted by the lack of private or corporate interest in funding university research. I received a five-year award that provided me with up to $37,500 per year in matching funds for every dollar I obtained from private and corporate sources for research. I had thought that it would be easy, with the help of our fund-raising staff, to obtain this matching money. Somehow, I had the idea that there was a sea of people who loved science and scientific research and was just waiting for an opportunity to support someone like me.

But when I went to talk to our fund-raisers, they were unusually blunt. They informed me that about a dozen other professors had received this award at my university over the previous decade. However, it was rare for any of the previous award winners to find matching money from private or corporate sources. Unless I had my own contacts with a private or corporate source (and that source was not already "taken" by someone else at the university), my chances of obtaining matching money were negligible.

At the time, I was absurdly indignant about this situation. I thought that the university fund-raisers were not performing their job properly. I also thought that if it was true that neither the public nor the corporate world were interested in furthering research such as mine, then there must be something horribly wrong with the nation. I don't think that my attitude was that far from the academic norm. Both in terms of research and funds for instruction, there is, unfortunately, a strong sense of entitlement in universities today. After forty years of generous federal funding, administration and faculty have come to expect continued growth and support.

Now I feel that if the private and corporate world does not find enough value in research to provide funding, then so be it. I can't point to a single event that caused this change in viewpoint. It was the result of an accumulation of observations in which it was clear that the benefits of the research culture spawned by the Golden Age came with some serious problems. The chief problem was that it caused us to discount our efforts toward undergraduate education.

My own view is that the public's and corporate world's focus on undergraduate education over research is a good thing. Ultimately it may cause us to better balance our teaching and research missions. If in the coming decades, government support of university research continues to be flat and private and corporate support continues to grow dramatically, we may as a result see a reorientation in focus. Since the public is far more inclined to support teaching than research, it may through its collective pocketbook change the face of the American research university. Under such a scenario, the importance of undergraduate instruction at these universities will likely resurface. Such a change would be welcomed by many, myself included.

12

You've Got to Believe

I have a friend from another department who fled Cuba as a child, and whose views on communism and, in particular, Fidel Castro are highly negative. I sympathize with his opinion. My parents were refugees as well. My mother spent much of World War II imprisoned in a Soviet work camp and my father defected from the Polish Red Army (used as cannon fodder by Stalin) after the war. But liberal to leftist politics dominate universities, and Castro's Cuba, despite its well-known deficiencies, remains a darling of many academics. When my friend told me that Duke was developing a student program in Cuba (our students would take classes in Cuba for six weeks during the summer), I commiserated.

My friend made his case against the Cuba program to the university administration. He knew that he had no chance of changing their minds, but felt a duty to lay out the obvious problems. He noted that we have no trade with Cuba. We have no diplomatic relations with Cuba and an oppressive dictator runs it. They have poor medical facilities and access to medicine is limited. Should one of our students become seriously ill, we would likely have to fly him or her back home. Serious problems could ensue should any student get a crazy idea to travel around the countryside and end up at a place that is off limits to foreigners.

Despite its potential negative repercussions, the program was approved by the international programs committee, and was given its final approval by our president. It was held for the first time in the summer of 1998. Looking at the course titles, it is not clear whether our students received education or political propaganda. When I asked an administrator about the motivation for creating this program, he smiled and said, "We want to send an olive branch to Cuba." Gee, I thought that we were a university, not a part of the State Department.

My own view is that if Castro were to abandon the pretense of being a communist and own up to being just another nasty dictator, my university would find Cuba far less alluring, and conversely, the United States would likely reestablish trade. The prospect of Castro making such an announcement, however, is unlikely. I may not think that the embargo is accomplishing anything positive, but it is official U.S. policy. I don't understand why my university views it necessary to thumb its nose at a governmental policy that has been in place for forty years. I can assure you that a proposal to develop a Duke educational program in Iran as an olive branch would go absolutely nowhere in the university. I expect U.S. sanctions against Cuba to loosen when Castro fades from the political scene, but we haven't quite reached that stage. There is no real rationale for our Cuba program, but it is consistent with the tendency of universities to become mired in liberal and leftist political advocacy.

University faculty across the nation moved strongly to the left sometime in the late 1960s and early 1970s and have stayed that way ever since. This was particularly true of those in the social sciences and humanities. The broad political shift of faculty in colleges and universities was a phenomenon related to the Vietnam War. The war protest movement was centered on college campuses. It served to radicalize a large segment of the student body, a significant number of whom went on to get their Ph.D.s and eventually become professors (they are now old enough to be deans and presidents).

Many of my political views are liberal. From affirmative action to the plight of the poor, my view is that government should play a role in ensuring social and economic justice. Given that the political leanings of most universities are largely in accord with my own, I guess that I should be pleased. But I am not happy with the way our leftist political slant has influenced university instruction and decision making. I also view as wholly inappropriate the small countermovement of the political right to encourage some universities and colleges (mostly those with a strong religious focus) to promote a conservative bias. Universities are in the education and research business. They are supposed to try their damnedest to leave their political biases out of both of these tasks. At stake is more than just the possibility of looking foolish. When universities mix teaching and research with political advocacy, they damage their credibility as institutions of responsible education and research.

Because we have a nearly monochromatic political view and one that is largely out of touch with the public, our actions, like the development of our student program in Cuba, sometimes appear ridiculous to the outside world. The problem is that universities follow academic political fashion, and when they develop new programs, public opinion has little influence. Rather, the focus is on impressing other universities. Universities can look somewhat like a Parisian fashion show to the public. Some of the programs they develop are beautiful and others seem workmanlike. But every once in a while they parade a program before the public that draws laughter and catcalls.

This tendency to look foolish is particularly true in the humanities. The Golden Age may have been a boon to those in the social and natural sciences, but for the humanities, the Golden Age was a mixed blessing. Universities were financially so well off that in a version of trickle-down economics (the only example that I have ever witnessed where such a process has worked), humanities departments received some of the bounty from federal support of the sciences. With this spillover, teaching loads for humanities faculty were, like those in the sciences, decreased to allow them more time to research. They also received small amounts of research funding of their own from federal and state agencies, and private foundations.

But the Golden Age also seemed to bring a loss of purpose to much of the humanities. Before World War II, the humanities were the heart and soul of universities, and humanities fields typically constituted the plurality of faculty and student majors. With the rise of research during the Cold War, however, science and engineering faculty growth outpaced growth in the humanities. At the same time, undergraduate students began to lose interest in the humanities, and by the end of the Golden Ages less than one-seventh of all undergraduates nationwide were choosing humanities fields for their major. Students were, instead, choosing more practical fields of study to increase their chances of finding a good job. By the 1980s, the political power of humanities faculty on campuses had weakened tremendously.

The humanities underwent a profound intellectual shift toward the end of the Golden Age that perhaps was related to their diminished importance. The seventies and eighties saw the rise to prominence of "cultural studies" within the humanities community, a broad area of study where the best of art, literature, and civilization is mixed with the mundane and trashy products of the present and the past. Some faculty members began to study comic books and television shows with the same level of seriousness they applied to Shakespeare. Some created classes with such titles as "Melodrama and Soap Opera" (a course that is taught at my university). To a significant portion of the humanities community, cultural studies was viewed as intellectually liberating and cutting edge. In contrast the common reaction of parents and legislators to the rise of cultural studies was that professors were goofing off and teaching their children nonsense. Traditional humanities scholars, who believed in studying only the lofty aspects of civilization, engaged in political battles with the cultural studies community, but they were frequently on the losing end of these struggles.

In the 1980s, my university made a conscious effort to increase its academic visibility in the humanities, allotting several million dollars toward this effort. With the lure of high salaries, we hired a number of prominent humanities faculty members away from other universities, many of whom were at the forefront of cultural studies. We have a few Marxists advocating political and economic change, and in greater abundance we have cultural theorists of the generic left whose research focuses on gender, racial, ethnic and class bias in art, literature, and society. Of course, we also have faculty members who espouse unbridled free market capitalism, but any argument that our hiring over the past two decades has allowed us to achieve political neutrality through equal representation of extremes rings hollow. Our leftist faculty far outnumber those on the right.

The presence of cultural studies gurus is a blessing or a curse depending on your view. The blessing is that they are highly regarded in the humanities and give my university national prominence among humanities academicians. Until a few of our more prominent faculty members in cultural studies left for other universities in the summer of 1998, we were probably the best university in cultural studies in the nation. Being number one in anything is almost always a plus. The curse is that the public finds much of the teaching and research in cultural studies nationwide to be pure hokum. Cultural studies is an easy target for the media to attack, and it is particularly easy because researchers in this field gleefully ignore public criticism.

Our cutting-edge humanities faculty have been pilloried in bestselling books, on television, and in major newspapers and magazines. For non-humanities faculty, it is painful to learn from the popular media about English professors who have students read Louis L'Amour for college credit or condemn historically important works of art and literature on political grounds (for example, Shakespeare's plays promote sexism and oppression of the lower classes). However, faculty outside the humanities generally don't understand cultural studies well enough to fully evaluate the validity of these criticisms, and are too busy to give them much thought. We know that cultural studies is a politically biased field, but political bias is so pervasive in academia that singling out one area of study seems arbitrary. Also, there is a tendency among faculty to possess faith that their colleagues are being responsible researchers and educators, and to discount media criticism.

Occasionally, however, an incident occurs that cannot be easily dismissed and casts serious doubt on the worth of cultural studies. For example in 1996, Duke University Press published a parody of cultural studies written by Alan Sokal, a New York University professor. The parody appeared in the press's cultural studies journal *Social Text*. A hoax full of intentional gibberish, it was so cleverly written that the editors did not recognize it as a parody and unknowingly and embarrassingly published it as a legitimate article.

Sokal, a professor of physics, is that rare bird in the sciences, a professor with a strong leftist political bent. For him, the problem with cultural studies researchers was that instead of effectively advancing social change, they spent their time writing nonsense. In his view, cultural studies research, particularly research that purported to study science culture (how scientists work and the value of science), was without merit. In his send-up of such research, "Transgressing the Boundaries: Toward a Transformative Hermeneutics of Quantum Gravity," he pandered to the cultural studies crowd by "arguing" that it "has thus become increasingly apparent that physical 'reality,' no less than social 'reality,' is at bottom a social and linguistic construct." He found fault with the idea "that there exists an external world, whose properties are independent of any individual human being and indeed of humanity as a whole." He used science analogies that sounded lofty but were completely inappropriate and butchered basic scientific principles. What he wrote made absolutely no sense, but making sense was clearly not a condition for publication in the cultural studies journal *Social Text*.

In a subsequent article published in *Lingua Franca*, Sokal explained why he had carried out his hoax. (His parody must have taken an incredible effort and amount of time to write. It is one thing to turn out just plain gibberish. It is quite another to turn out gibberish that looks and reads like something that has real content.) He stated:

The targets of my critique have by now become a self-perpetuating academic subculture that typically ignores (or disdains) reasoned criticism from the outside. . . . I offered the *Social Text* editors an opportunity to demonstrate their intellectual rigor. Did they meet the test? I don't think so.

The news of Sokal's parody traveled around the academic world and was featured throughout the media. My university's administration chose to make no official statement about the incident. When pressed by reporters, administrators tried to distance Duke University from the hoax by noting that the principal editors of the journal *Social Text* were not faculty members. They tried their best to ignore the connection between the university press and the university. Cultural studies faculty across the nation expressed outrage at the hoax. One of our more prominent faculty members wrote an editorial in the *New York Times* that condemned Alan Sokal, and with logic as poor as that contained in Sokal's original parody, attempted to justify the value of cultural studies research of the sciences.

The net effect of the parody on the fate of our cultural studies community—and those communities elsewhere—has probably been negligible. The university still avidly supports cultural studies research. In fact, when several of our cultural studies faculty announced in the spring and summer of 1998 that they were leaving for greener pastures, our administration quickly appointed a committee to help hire replacements. Duke's academic press still publishes the journal *Social Text*. Cultural studies researchers have not changed their focus or methodology. What the parody did was to highlight the difference between academic perception and public perception. Our English department, ranked as one of the top ten departments in the country by its peers in 1993 (largely on the basis of its strength in cultural studies), still is highly regarded in the academic community. In marked contrast to the importance attached to cultural studies in academia, cultural theorists continue to attract the ire of both the media and tuition-paying parent.

I used to think that those in the sciences and engineering were completely untouched by the political bias in universities. Oh, I heard about the craziness that went on in other areas of the university secondhand. Students have told me that cultural studies classes sometimes degraded into a game where you did well by writing papers that emphasized the horrors of western capitalistic society. I suppose that there is value in learning how to suck up. I am also aware of the equally irresponsible behavior of some faculty members in economic departments who require students to unquestioningly embrace the virtue of free markets and to deride government economic intervention. But in science departments, there has been little in the way of a social and political agenda. We have managed to avoid much of the political advocacy and general lunacy that goes on in portions of the humanities and social sciences. Politics just is not an emotional issue for many faculty in the sciences.

For scientists and engineers, the exposure to the incredible world of political correctness and bias is most often found by being a member of university committees. Like my friend who watched helplessly while we created an educational program in Cuba, I too have been on the lopsided losing end of committee votes. Being on a university committee that meets once a month and watching an occasional dumb idea or project get approved because it is politically of the correct flavor is the price I pay for being an academian. While it is irritating and sometimes galling to be part of this process, the issues that I've voted on and lost have had little impact on me personally.

But because some of my research is in the arena of environmental science, I have found that I am highly affected by political bias in universities. As a hydrologist, I've performed research on a wide variety of environmental topics where water plays a role. I teach classes in environmental science. I am strongly interested in environmental issues, but I am only a mild advocate for environmental preservation.

My own view is that great strides have already been made in the environment. The overall environmental health of our nation has been markedly improved since I was a child. Images of large-scale industrial pollution such as the burning Cuyahoga River in Cleveland, and the toxic Love Canal in New York are largely a distant memory. However, the environmental movement has a tendency to undersell the gains that have been made, and I find environmental lobbyist's calls for additional legislation in the 1990s to be frequently unrealistic and without scientific basis. In contrast, many environmental scientists are ardent supporters of environmental lobbying organizations. It's predictable that this is so. After all, why work on environmental problems unless you have a deep seated love of the natural world?

While personal political bias is common among those who work on environmental issues, it would seem essential that university-based environmental researchers and educators make the effort to remove bias from their work. However, some environmental academicians argue that, since we are all politically motivated creatures and environmental issues are loaded with political concerns, it is not possible to be politically neutral in research and teaching. I disagree and know from personal observation that most of my faculty colleagues in the environmental arena work hard to be unbiased. Unfortunately, there are also those who make little or no effort to uncouple their political views from their research and teaching. Researchers in universities sometimes overemphasize the negative repercussions of human-induced change on the environment. (To be fair, I've seen a greater degree of bias in some government agencies, most notably in the Environmental Protection Agency.) The tendency to proselytize for environmental causes and dwell on doomsday scenarios of the future of our environment is more pronounced in the classroom than it is in research.

To promote environmental causes, a professor does not have to lie outright. In the classroom, the professor can simply omit persuasive arguments from the opposition. In research, he or she excludes procedures in the design of experiments that would likely provide conflicting data. Privately, colleagues have argued to me that preservationist bias is essential to counter academicians and Ph.D. consultants who receive funding from corporate interests and downplay environmental impact.

The preservationist slant of university-based environmental research and instruction seems to be better tolerated by the public than the leftist bias found in the humanities and social sciences (leftists don't have sole dominion over environmentalist advocacy). It is still inappropriate and negatively affects our credibility. Regardless, universities tend to be such bastions of environmentalism that these biases in research and the classroom are accepted and sometimes openly encouraged. Also, independent of political bias at universities, there is a live and let live attitude amongst faculty and administrators concerning research. As long as a professor receives research funding, colleagues and deans seem to care little whether or not the research is being used for political advocacy.

I have been both aided and hurt by the promotion of environmentalism on my campus. It's been nice to see money and resources come my way. But it would be better if the money was associated with programs that made sense. Similar to our Cuba program and cultural studies research, programs that have environmental themes are politically popular in universities and when universities create and develop them they tend to suspend critical judgment.

For example in 1990, the year of my arrival, the university created a School of the Environment. By doing so, it was following a small and problematic nationwide trend begun in the 1970s. By 1990, there were about twenty environmental studies programs in research universities across the country with natural and social scientists from diverse fields loosely organized under a broad environmental umbrella. The ostensible motivation for creating these broad-based programs is that the solution to environmental problems requires people working together from many different areas of expertise.

The need for such "interdisciplinary" effort in environmental science is real. However, interdisciplinary programs of any kind run counter to the traditional approach to organizing academic elements of universities, and as a result, their creation requires careful planning. Environmental programs, unfortunately, generally have not been developed with a great deal of forethought. Instead, their creation appears to have been predominately motivated by an emotional need to have a campus presence in environmental advocacy. These programs have not been successful academically, and their potential value to universities and society has gone largely unfulfilled.

Because their creation is largely politically motivated, environmental studies programs tend to preferentially attract faculty who let political bias influence their work. This tendency undermines program credibility, and makes it difficult to attract faculty who are honest in their research and teaching. Many excellent faculty members in the environmental field avoid participation in environmental studies programs. As a result, these programs generally have had little intellectual impact. Breakthrough research in environmental science continues to be dominated by faculty members from biology, chemistry, physics and geology departments. The political agenda of these environmental programs also leads to deficiencies in undergraduate education. While popular with students, the education is long on pandering to environmental causes and short on giving students training in the fundamentals necessary to solve environmental problems.

It was clear from the outset that our School of the Environment would be subject to the same deficiencies as other environmental programs. The final summary document of the task force that put together the school was oddly lacking in substance. Instead, it read like a tract written by environmental lobbyists. The second page of the document was devoted to a *New Yorker* cartoon that depicts a well-dressed woman talking to a tall, fat man at an upscale cocktail party. "I'm an environmentalist," she says. "I suppose that frightens you." The rest of the document was awash with praises of the university's expertise in environmental studies. It was full of passion, but lacked a coherent strategy for success.

The rationale for creating the school was partly based on wishful thinking that environmental research and employment would grow dramatically in the 1990s. Large increases in federal spending to preserve wildlife, and improve water and air quality were expected. It was also assumed that research funding for emerging large-scale environmental problems such as global warming would grow tremendously. These expectations proved to be false, and for several years, our new environmental school struggled financially.

The slowdown in the environmental field was predictable. By 1990, it was fairly clear that after twenty years of rapid growth in environmental regulation, the nation was reluctant to spend more money on clean up and preservation. Unlike my own reluctance to support more environmental regulation, the mainstream public's opposition seemed to have little to do with any perception of the frequent lack of scientific basis to proposed new regulations. For the public, the issue was mostly economics. They did not want existing environmental regulations repealed, and many wished that the money already allocated could be used more efficiently, but their attitude toward the environment had evolved to something akin to their attitude toward public education. They were concerned, but not concerned enough to open their wallets. While universities like mine expected dramatic increases in environmental funding and employment, the 1990s were a decade when the business side of the environment underwent maturation.

There are, however, advantages to being part of a program with high politically correct visibility. It was just not acceptable to our university's administration to allow the school to fail due to a lack of sound initial planning. The university covered million-dollar expenses that the school could not afford. Finally, a loyal and generous alumnus donated $20 million to the school, putting it on much sounder financial footing.

While the school was financially stable in ensuing years, it continued to struggle in other ways and suffered from the typical credibility problems that hound environmental programs. In 1997, Duke's environmental school underwent reorganization in an attempt to prop up its image. Rather than continue to have an amorphous structure, the school began to be organized into divisions. Geology, the department that hired me, was moved into the school and renamed the Division of Earth and Ocean Sciences. Four other divisions were in planning as of 1998.

Will universities like mine eventually succeed in their goal to develop centers of nationally recognized excellence in environmental research and education? Obviously, I am a skeptical participant in this effort, but I'm hoping that they do. Since these programs are devoted to a politically correct topic, I expect that they will continue to be internally financially supported, and money usually can cure a lot of problems. There is certainly a societal need to have academically credible, broad-based environmental programs within universities. We just haven't been particularly careful in creating these programs, and they will continue to be academically unsuccessful without critical evaluation by faculty members and administrators. Programs whose creation and support are motivated by a desire to be politically fashionable within the academic community tend not to undergo scrutiny. Rather they are allowed to continue unchecked because of a faith in their inherent goodness.

For me, whatever tyranny exists in political correctness does not reside in the classroom. I don't feel that my freedom as a teacher has been hindered in any way over my time as a professor. Of course, I can't employ humor that is religious or sexually based (while growing up in a neighborhood that was half-Jewish and half-Catholic, I developed a fondness for rabbi-priest jokes), but I still manage to find more than enough jokes to tell. Rather, the problem is that political correctness has forced us to collectively suspend our judgment on decisions made by faculty and administration that have clear political bias. It is acceptable for faculty to be critical of many aspects of a university. But when it comes to programs like those in Cuba, cultural studies, and environmental studies, we are supposed to ignore their faults. Programs such as these are given a status that many assign to their religions. We can't question or doubt, and must blindly believe.

13

The Fifty Percent Solution

I was seated in an audience that consisted almost entirely of women, most of whom were graduate students in the sciences. The topic of the lecture, "Paths to Success, Women as Science Professors," only mildly interested me, but a friend of mine convinced me to attend. The crowd was fairly large for an end of the week seminar. The lecturer was a female faculty member from another prominent university. She was short and slight of build, energetic and upbeat. She spent the first part of her lecture recounting her own considerable successes. A full professor in her early fifties, she was a part of the very small contingent of female professors who began their careers in the heyday of the Golden Age. She was proud that she had balanced both career and family successfully. Her two children had grown up to be successful in their own right. When she was a young mother and untenured faculty member, she did not have the time to spend the same number of hours performing research as her colleagues. But this, she insisted, did not put her at a disadvantage. During her tenure evaluation, her tenured colleagues noted that she had managed to make extremely efficient use of her time. "It's not the number of hours that you work, it's what you do with your hours that counts," she said.

I am sure that "I did it and you can too," speeches can have a positive influence, but part of me wanted to hear the full story. Her personal history seemed altogether too blithe and free of obstacles. Her colleagues were too full of understanding and support. They seemed altogether too eager to accept her as a full-fledged colleague. Universities are, like the corporate world, dominated by men and women often have a difficult time being taken seriously. The time period when the lecturer was an untenured faculty member was an awful time for women in universities, particularly for those in the sciences. It may have been the Golden Age in terms of university growth, but for women the era was far from golden. Their applications to graduate school were often ignored. Their searches for academic jobs were often stymied. And once they obtained jobs, their prospects for achieving tenure were worse than those of their male counterparts. Life for female faculty has improved considerably since the Golden Age, but they are nowhere near on equal footing. Surely, this lecturer must have faced some adversity related to her sex along the way and there would be some motivational value in these students hearing how she conquered these problems.

There was a part of me, however, that enjoyed hearing her unfettered tale of success. Female faculty who came up during the Golden Age were pioneers and as such tend to be a hardscrabble lot. It was refreshing to hear a history so upbeat and devoid of rancor. Maybe she was lucky and did not have to face too many hardships.

But when she got past her personal history, her lecture took a different turn. "We do not have a critical mass of female science faculty in this country," she said. "Without a critical mass it is difficult for women faculty and graduate students to feel like they belong." She briefly discussed the central problem with achieving such critical mass nationwide. Women do not enroll in the sciences in numbers equal to men as undergraduates. As a percentage, a smaller number of them go on to graduate school. When they attend graduate school, they do not go on to first-tier academic jobs in equal numbers. In academic jargon, this is known as "a pipeline problem." Along both segments of the pipeline, undergraduate studies and graduate studies, we lacked enough women to assure a critical mass of faculty.

When most people talk about lack of women, they stress that we must find ways to fill the pipeline and encourage women to enter the sciences as undergraduates and continue their studies through their Ph.D. But the lecturer chose to only touch briefly on this potential avenue for change. Instead, she advocated a new approach to achieving critical mass. Since there were not enough faculty members to go around, we should concentrate female science faculty at a few universities. The benefit of doing this would be twofold. The women faculty in these universities would be in a more hospitable environment and as a result could be more productive. These institutions would also become magnet centers for females to attend graduate school in the sciences. The students would have many role models and be more likely to choose an academic career. She noted that her university was playing an ad hoc role in concentrating female faculty by actively recruiting tenured female faculty members from other universities. "We have already achieved considerable recruiting success," she said.

I looked around and noted the older female faces in the crowd. We have some excellent female faculty in the sciences. Perhaps our lecturer was soliciting one or two of them during her visit. I didn't want them to be raided by another university. Personally, they might benefit from such a move but their presence here had a positive impact. "How much has she offered you?" I whispered to my friend, a tenured female faculty member with whom I had come to the lecture.

My friend gave me a wry smile. "I wish," she said.

Walking back from the talk, we discussed the virtue of grouping successful women faculty into a few universities. My friend dismissed the idea. "It's too artificial. We just need more women faculty throughout academia," she said. When I asked her how many more faculty were needed she said that it was important that departments have a critical mass of women faculty. "We need to find a way for departments to have several women faculty, not just one or two." She echoed the sentiments of the speaker when she said, "Otherwise, female academicians will continue to feel isolated."

Several years later, I found myself having dinner with five female faculty members from around the country. Three had started at the end of the Golden Age and were already tenured. Two were post-Golden Agers whose careers were going well. Aside from the typical nervousness and anxiety concerning tenure shown by the two youngest professors, they were a confident and assured bunch, part of the small wave of female faculty who followed the pioneers of the 1950s to 1970s. They came up during a time when female faculty members were a significant instead of a negligible minority. In general, they had not suffered nearly as many indignities and as much discrimination. But women professors are still a minority on college campuses, particularly at the elite universities. It is telling that the only higher education campuses where female faculty approach equality in numbers are two-year community colleges. As one looks higher and higher on the prestige (and salary) ladder, the percentages drop significantly. At the elite private universities, female faculty represent about 20 percent of the faculty and 5 percent in the physical sciences and engineering. Two of the women at the dinner table were the sole females in their departments. I thought of myself in this group. As the sole male, I was a distinct minority. Did I feel comfortable surrounded by these women? In many ways the answer is yes, but ultimately I felt I had to work harder at conversation than I usually do around men. In some ways, it reminded me of having dinner with people from another culture. I had to listen more carefully and measure my words. These women had to adapt in a similar way in their jobs. Only they didn't have to do it for only a couple of hours. They had to do it for their entire careers.

I thought back to the lecture that I had attended a few years back on women as science professors. I wanted to hear a more complete version of the life of women in academia so I asked my dinner companions what principal problems they had encountered as female faculty members. The central issue was one of acceptance and respect. "We're not listened to. Sometimes it's as if I'm invisible. I've said things at faculty meetings that are completely ignored. A few minutes later a male faculty member will repeat what I've said and it gets all kinds of attention."

Two of them nodded their heads in agreement. "When I go with my husband to university functions, it's commonly assumed that my husband is the professor," another said.

"Outside our central departmental office," she continued, "we have photos of faculty and staff. The faculty are identified by the title 'Doctor' and then their last name. When my photo was first placed on the bulletin board, I was identified simply as Sarah Anne. No last name and no doctor. When my mother noticed this during a visit, I had it changed.

"And it isn't just the faculty and staff. It's some of the undergraduate students as well. They refer to me as Mrs. Maple. My colleagues are all called Professor and Doctor by the undergraduates."

"The students expect you to be nicer and more lenient," another said. "Especially with quantitative material. Somehow they expect that because I'm a female they won't have to deal with equations. And when I get quantitative, they become very resentful.

"If a male professor is tough, undergraduates take it in stride and say, 'Well I guess that's just the way he is.' If I'm tough, well then I'm just a bitch."

"That goes for my colleagues as well," another said. "I'm expected to not make too much of a fuss. I'm a bitch if I put my foot down on an issue."

"My chairman once told me that I don't promote myself enough," another said. "That week I found an article in a magazine about differences between men and women that among other things mentioned that women aren't inclined toward self-promotion. I cut out the article and put it in his box. When I asked him what he thought about it, he said, 'See I was right. You need to promote yourself more.' "

What they said agreed with what I've observed as a student and professor. Female faculty members are often not given the attention and respect they deserve by both colleagues and students. Partly it's physical. They're smaller. They have higher voices. Mostly it's sexual. Women are viewed by many men as sexual objects first. They are not expected to be intellectual authorities or command political power. They have to prove themselves in ways men don't. Even then, many men refuse to take them seriously. If I had been more comfortable with these women, I would have told them about a female friend of mine who published an article in a prestigious journal. At a faculty meeting, a colleague of hers said, "I saw your article. Nice job." Before she could feel too proud, he said, "I was surprised that they didn't include a centerfold."

I asked them what they viewed as the stereotypes of women faculty. "Well there are the obvious ones. For instance, we're supposedly too emotional."

"We focus on students and instruction to the detriment of our research," said another.

"We aren't aggressive enough."

"Do you think that many believe you were hired only because you are female?" I asked this because it's a fairly common conversation among male graduate students that they will never find a job because universities preferentially hire women.

"Sure, it's in the background," one said.

"What about the role model business?" I asked. I noted that women faculty members on my campus, particularly in the sciences, spend more than their fair share of time on public display at university functions and on university committees. Personally, I would chafe at performing so many activities not directly related to teaching or research. But my dinner colleagues were largely nonplussed with the extra workload.

"I like being involved in campus issues. But when I'm asked to be on a committee, I try to find out if I'm being asked simply because they want a token woman," one said.

"All the requests can be a good thing too," another said. "You can use it to your advantage. You get exposed to many aspects of the university. The same holds for being asked to be on national governmental panels."

"What about the issue of critical mass?" I asked. "How many faculty members need to be female before women can be fully accepted?" As I asked this question, I had my own answer in mind. I was guessing that they were going to say something on the order of 25 percent. There are typically twenty to thirty faculty members in a university department in this country. If these departments each had five or six female faculty, I thought, women would constitute a significant percentage of the professorate. But my opinion was not close to theirs.

"We need 50 percent women faculty," one said. "If we had 50 percent we would have acceptance." No one disagreed with this assessment. I tried bargaining with them. "How about 20 percent?" I asked. But there were no takers of my offer. I thought about this issue for a few minutes and decided that they were right. Women will not obtain full acceptance in universities unless they represent half of the faculty population and half of the upper echelon administrative population. The only problem with this goal is that such a percentage is not likely attainable in the near or distant future.

Looking at research universities across the nation, women in the 1990s comprise about 25 percent of all tenure track (those eligible for tenure and those who have tenure) faculty. The distribution of these 25 percent is not evenly distributed across college campuses. Humanities departments get the lion's share, followed by social science departments and natural science departments.

These percentages represent dramatic improvements over those in 1970 when women comprised less than 10 percent of all tenure track faculty at research universities. The early 1970s were a watershed time period for women faculty. Federal laws banning discriminatory hiring on the basis of sex and affirmative action forced universities to hire women faculty in larger numbers. But by the 1980s, rates of hiring of women faculty began to plateau and the percentage of faculty who are female has not dramatically changed since the late 1980s, particularly at top notch universities. For example, the percentage of new hires that are female at the University of Wisconsin has shown no upward trend over the last decade and remains at around 31 percent. Yet, women constitute the majority of undergraduates on the nation's campuses and receive 39 percent of all Ph.D.s, women are just not competing evenly with males. On a percentage basis, males are more likely to go on for a Ph.D., and even more likely to get plum academic jobs. Barring any dramatic changes in universities or in social mores it is clear that my female colleagues will always be a minority on major college campuses. In the sciences and engineering they will continue to be vastly under-represented across academia. Women already in academia may want half the new jobs filled with women and I may agree, but it just isn't happening.

Why haven't percentages increased significantly over the past decade? An obvious reason is that despite federal laws, sexual discrimination still occurs at universities. Hiring of faculty members begins in departments. In some departments at universities, particularly those where females are scarce, new female applicants are just not given a fair shake. The sexual discrimination doesn't have to be overt. In this day and age where almost every new job opening has in excess of 100 applicants, many of whom are worthy of being hired, the differences between the top candidates and the also-rans are usually very subtle. When departments hire they do so not only on the basis of academic promise, but just like the business world, most worry about whether the candidate is a good fit in terms of personality. They want someone that they can feel comfortable with at departmental meetings and at the occasional departmental lunches. If the men don't particularly feel comfortable around women in the workplace it is easy for them to find subtle reasons to ensure that females do not rank at the top of the applicant pool.

I haven't been exposed to departments like this myself and I'm going to optimistically assume that the number of departments where sexual discrimination occurs in the hiring process is small. In my experience of hiring new faculty, women haven't been hired simply because they haven't applied in the numbers one might expect. Women constitute over 30 percent of the graduate students and 20 percent of the doctoral degrees awarded in my area of study. Yet when we search for new faculty in my department, women constitute less than 10 percent of all applicants. My department has only two female faculty members, but we would have more if women would apply for the jobs that we have to offer. For some reason, women are not finding our academic jobs to be all that attractive.

Why aren't women applying? At face value the jobs that we have offered are among the more prestigious in the country. If you want to pursue an academic career, my university is about as good as can be found. But when you scratch the surface of jobs at top-notch research universities, particularly those in the sciences and engineering, you quickly see that there is a downside to them. Unless you are a true genius (and I've only met two or three professors in my life who I would classify as geniuses), these jobs generally require a singular devotion to career, particularly during the untenured years. The work requirements, which principally consist of research, have actually increased since the Golden Age. For someone with a broad perspective on what is important in life, such jobs hold little appeal. The time when the work requirements are greatest is also the time period most likely for women to have children. I will generalize and say that women tend to have broader and healthier life goals than men. They don't find such a relatively monochromatic way of living desirable.

The childbearing issue is clearly important for women deciding on an academic career. A woman who has advised many female graduate students wrote to me, "When I look at women graduate students that I have trained as well as some women friends who received their Ph.D., it seems to me that existing or anticipated family obligations are among the biggest factors keeping them from pursuing academic positions. Even though more women faculty are managing to have both careers and families, I think it still looks like a pretty tough road ahead for women who either had children during graduate school or hope to have them before their biological clock stops ticking. . . . I know of several cases in which women have decided to leave academia because of the difficulty of balancing the needs of children with the demands of an academic career. There are relatively few 'househusbands' compared to the wives of male faculty who are able to take several (or many) years off from a career to provide for young children."

Then there is the job itself. Life in the ivory tower doesn't appeal to most people in general and women in particular. At research universities, you're rewarded almost entirely for your research. The standard path to success is to work on a well-defined (and to the public esoteric) topic for many years at a time. It pays to not only do your work well, but to engage in self-promotion at national meetings and other gatherings. It is important that you publicly receive credit for your work. When you work on a group project, your colleagues need to know or at least must be made to think that you are the brains behind the effort. True social interaction is often rare and the connection to society is indirect. The rewards are based on a system that strongly values individual achievement as recognized by fellow academicians at other universities. The one avenue of the job with direct societal involvement, teaching, is not considered to be all that important by the powers that be and by most fellow faculty.

I will generalize again and state that women like working in teams on issues that have direct relevance to society. Our ivory tower jobs tend to attract those who want to be alone in the spotlight, yet seen only by other academics. To quote one of the women with whom I had dinner, "I like to work collaboratively on problems. I like to work on problems that have environmental relevance. The definition of how one should work and what one should work on to be successful is pretty limited in academia. Universities emphasize the success of the individual and the importance of NSF funding."

Finally, there are the experiences students have when they are graduate students that give them a window into university life. Graduate school can be the equivalent of an extended boot camp. It is not all that unusual to have professors lord over their graduate students imperiously. They can be cruel and treat their students, who are typically in their mid to late twenties, as if they were children. Graduate students receive a stipend for their research that is typically around minimum wage and many faculty rarely pat them on the back for their efforts. For women more so than men, this approach to education and training seems pointless. Personally, I don't see the rationale behind this approach either, although I've witnessed it fairly frequently. If you have been working as a Ph.D. student for four to six years under conditions where you have been treated with little respect and are sometimes humiliated outright, you may not have much desire to be anywhere close to a university once you receive a degree.

Given these less than desirable characteristics, the reader might ask why anyone, male or female, would want such a job at all? The answer is that, despite its obvious detractions, our current mode of operation is extremely appealing to those who, like myself, have a strong individualistic streak. As a professor, I have an incredible amount of freedom on the job. Ultimately the decisions on what research avenues I choose are made by myself. I have considerable freedom to choose what material I put in my courses and how I teach my courses. For me, the most negative aspect of the job is the workload, which keeps me away from my family more than I would like. However, in talking to friends with well-paying jobs outside of the university, I know that my workload complaint is definitely not unique to academia. Also, I've talked to a couple of professors who are incredibly efficient and don't find the workload to be all that harsh. "It's a myth that you have to spend so many hours working that you can't have a family," a female faculty member from another prominent university once told me. "When I talk to female students, I try to make them aware of this."

Despite the fact that women are underrepresented in the faculty and administrative workforce, the entrance of women into academia has influenced the workings of universities. In some ways, the influence has been significant. It is doubtful that our sexual harassment policies would be as extensive as they are were it not for the entrance of women into positions of authority. The presence of women professors has given universities a significant number of role models for female undergraduate and graduate students.

But most of the influence of women has been fairly subtle. We have created women's studies departments in many universities largely in a symbolic and token effort to recognize the importance of women. We have made some modest efforts to reward and encourage teaching across the country, and my guess is that this change partly came about because the new women faculty members were more concerned about instruction than the male faculty members they replaced. In our evaluation of candidates for tenure, we spend a little less time counting the number of research papers and books published by a candidate and a little more time examining the quality and impact of those papers and books. This change may have been partly the result of placing women on tenure committees and because women faculty members tend to be less obsessed in their own careers with publishing as much as possible.

It might be expected that women faculty members would lobby hard for wholesale changes in the way we do business. However, the numbers of women faculty members are too small to expect more than modest political clout, and most faculty, male or female, are often too busy with their careers to advocate for internal change. Plus the female faculty members we currently attract are those who either find little wrong with our current way of operation or at the very least have learned to tolerate it. As an example, I note that my university is one of a very few that has a woman as a president. Over her five years, we have continued on the same course followed by other universities. It is a course that evolved during the Golden Age, and although it is not particularly friendly to women as professors, we seem unable to change, fearful that our competitive standing relative to other universities will be harmed. In the process we and others have done little to make universities more attractive to women.

The reader would be wise to question how a male faculty member whose family life is more or less a throwback to the 1950s (my wife does not have a paying job, and spends a good deal of time with child rearing, family finances, and household organization) knows what women who might aspire to be academics want? The obvious answer is that I probably don't know very much. I will, however, have the temerity to sketch out some proposed changes. We have to create a university that is more community minded and less obsessed with the glory of the accomplishments of individual faculty members. In this different kind of university, careful teaching and interactions with students in general would be more than just encouraged. Promotions and raises in salaries would take into consideration significant involvement with students. The promotion and tenure process would have enough flexibility to allow both women and men to take extensive (not just one semester) leaves for having and rearing children. Universities would have day care facilities large enough to accommodate everyone who wishes access. Our education of graduate students would eschew combativeness for combativeness's sake. Research would also be more of a community experience and less effort would be placed on trying to differentiate between generals and soldiers on research projects.

These proposed changes are probably well off the mark of what is actually needed. I do note that my model of a female-friendly university is only partially in-line with what I think would be best for me. For example, I happen to like the way universities focus on the achievements of the individual.

When I asked the women with whom I had dinner what steps should be undertaken to improve the situation, one of them pointed to leadership and structure.

"We need to have more women in charge," she said. "The way that we work together extends beyond just how we communicate. It includes how we organize work, how we give credit, how we give feedback, and how we foster development. When I sit down and really think about it, there is almost nothing about the medieval structure of the university that I would ever have set up in my craziest dreams, had I been the one to organize the institution."

Another pointed to making the institution more attractive to those who are raising families, but also noted the difficulties of implementing such changes. "The problems of balancing family and career are not unique to academia, but the finite time period of the tenure track presents problems. Although universities have taken some good 'structural' steps, such as spousal hire policies and formal stopping of the tenure clock during a parental leave, the attitudes and expectations of faculty colleagues can't be changed so easily. I know of cases where women have decided not to take leave to which they were entitled because they worried that their colleagues would perceive them as less serious scientists. I also have seen evidence of this loss of respect when a woman faculty member takes time off with young children, so the fears are often justified. Having more women faculty who have found ways to have both a career and a family would make academia a more attractive option for my students as they complete their Ph.D. I think this change in climate has already begun, but there is still a long way to go."

Given the comments of my female colleagues, both their significant complaints and requests for structural change, it is clear that our current Golden Age model is far from ideal for women faculty. Their views can't be very far off from the mainstream. The proof is in the numbers. If the current way we operate is so great, then where are the women?

14

Making Adjustments

My family and I were just getting started. Our furniture and boxes had arrived from California after a delay of a few days and, after sleeping on the floor for a week, we fully appreciated the worth of having beds. We spent two days unpacking our stuff and settling into our new home. Then I put the boxes that contained my work materials into our car and headed off to the university. This would be my first day at work. I was anxious to get off to a good start and I had plenty to do. Classes would be starting in a week. I had to get some notes together to start my class. I was in the middle of organizing a scientific conference in my specialty area and I needed to deal with some logistical issues and firm up the speaker list for the conference. I had written to my department chair about a month before, telling him that this was the day I would arrive on campus. I was raring to go.

It was just a ten-minute drive from my home to my department's building. I had picked out an office during an earlier visit to my university. The building was undergoing major renovation, and the plan was that my office would be ready before I arrived. But there are usually delays in construction and this renovation was no exception. I had been told of these delays about a month before. So when I arrived with my boxes I didn't know where I would be temporarily housed. I went to my departmental office with some boxes on a dolly in tow.

The entire building was in the throes of remodeling. Dust was every-where. Holes in the plaster were common and the sound of a jack-hammer exploded in the background. Growing up in a family whose livelihood was the construction business, I was nonplussed. I went to the administrative assistant, said a brief hello, and asked for the key and directions to my office.

"I need to start unpacking," I said, turning my eyes toward my boxes.

"Those are your boxes?"

"Yes, they are. I've brought some of my stuff with me. There's more in my house. I'd like to unpack and get my office put together."

She paused for a while before she responded. "Let me call the chair." I waited while she talked to my chair, the head of the department. A longtime member of the faculty, he had taken the job one month before. The previous chair had hired me, but after twelve years in the position he was relieved to relinquish the task.

"It's just as I had thought," the administrative assistant said after talk-ing to my chair. "Unfortunately, we don't have an office for you yet."

"You don't?"

"No, we still have to move some students out first."

"How long will that take?"

"A few days at most."

"A few days?" I asked. "Where can I put my boxes in the mean-time?"

"We don't really have a place for you to put them."

"You mean I have to bring them back home?"

"Yes, I'm afraid you'll have to do that."

I was not a happy camper. Classes were starting in a week, and I had no place to unpack and get my work done. I called up my chair, partly because I couldn't believe that I had no office and partly to try to speed up the process.

I had met the new chair during my job interview. He seemed nice enough and when I thought of him as a chair, I wasn't worried. Mostly I wasn't fearful because I didn't know that a chair carried much weight. At the universities where I received my undergraduate degree and my Ph.D., the chair of the department was a position that was rotated every three years. Professors generally had no wish to be chairs. They didn't like paperwork and the act of balancing the needs of faculty with those of administration can be emotionally wearing. The chair was essentially forced to sacrifice research for a three-year period for the good of the department. Volunteers were not easy to come by. Hence, the rotation was mandatory.

Duke followed a much more hierarchical model and one that is common in private eastern universities. Its chairs were somewhat like middle managers in a corporation. The chair was chosen by the dean in consultation with the faculty and his or her role was to be in charge of almost every aspect of the department. He or she set salaries and raises, assigned classes and class times, hired support staff, and purchased departmental teaching and research equipment often with little input from fellow faculty. Faculty deferred to the chair. The chair deferred to the dean. The chain of command was extremely formal. It was not uncommon for a chair to aspire to move up in management and be a dean someday.

Almost immediately, I grew to hate this structure and viewed it as antiquated. It was appropriate for institutions whose success depended upon teamwork. Its origins can be found in the Bible, a structure God gave to Moses to organize his nation in the desert. In the modern research university teamwork is useful, but ultimately not all that important. (Besides it took the Jews forty years to cross a land the size of the state of Indiana. Perhaps a hierarchical organizational model isn't well suited for teamwork, either.) The universities with the loftiest reputations are those with the most outstanding scholars. Professors typically achieve their scholarly reputation through individual achievement. Somehow I assumed that all universities knew this and organized themselves in a way that got the most out of their faculty. I thought that the principal role of administrators was to facilitate the research and teaching of the faculty and the education of students. It took me a long while to realize that facilitators of any stripe were rare and I was bound to have a rocky beginning. Duke's organizational structure was more or less neutral in terms of its benefit to faculty. On the plus side I had a great deal of independence. On the minus side I had to find almost all of my opportunities—both big and small—on my own.

In response to my office problem, my chair said, "I really can't do anything for you, Stuart. You're just going to have to wait."

"I have classes in a week," I said.

"We all do, Stuart."

"I wrote to you a month ago that I was coming today."

"You can't always expect service," he said.

I expected an office and I thought my expectations were reasonable, but they were far off the mark. This fundamental mismatch between expectation and reality was one that occurred frequently in my first couple of years. I would have to make adjustments.

My initial problems with Duke were due in part to its location. I didn't understand southern culture at all. Unwittingly, I insulted people by being direct and forthright. Courtesy is very important in the South. There are codes of behavior you have to follow in discourse. For example, I didn't know that in the South it is extremely rude to simply ask for favors. You need to talk with someone about their family and shoot the bull for a few minutes before you can sneak in a request. Of course, like anywhere else, having some bigwig who can put in a request for you is even better.

For example, after a few days, I did receive an office. Now, I needed a phone. I had the departmental administrative assistant put in an order. A week passed and nothing happened. I then asked my department chair to help expedite matters, but he was busy with bigger issues than my phone. "You should be thankful you don't have a phone, Stuart," he said. "You don't have to deal with distractions." He was probably making a joke, but at the time I didn't think that his comment was funny.

I didn't have the necessary pull to speed things up. And my own efforts at getting a phone certainly did not help. Knowing what I know now, I would have walked into the office of the assistant to the assistant of the phone manager, shook his hand firmly and said something to the effect of, "Hello Mr. So-and-So, how are you doing today?"

"I'm doing alright, sir."

"Nice weather we're having."

"Yes, it is nice isn't it, sir."

"You don't have to call me sir. Just call me Stuart. Why the weather's been so nice my wife and I worked on the lawn all weekend, reseeding and stuff."

"Yeah, it was a good one to work outside, Stuart."

"Yeah, it was. Now, I know my wife has been trying to get hold of me all day. Something to do with our daughter's school."

"She in trouble?"

"I don't know. I hope not. She's a good kid. I hope that she's OK. She's just been a little sick lately is all. The doctors don't know exactly what is wrong, but they'll figure it out. I'm sure of it. It would sure help if I had a phone in my office though."

"You don't have a phone?"

"No, not yet. It would sure help me though, given that my wife is worried about our daughter. You think you could get one installed for me, Phil?" And I would have a phone in a day or two.

Instead, I waited seven weeks to receive a phone. And my efforts to expedite my phone request were disastrous. I walked into the office of the assistant to the assistant manager of telephones. I shook his hand

and launched right into my request. "Hello, my name is Stuart Rojstaczer," I said. "I am a new faculty member here. I've been here for more than a week and my office does not yet have a phone. I'd like to have one installed promptly."

"What did you say your name was, sir?"

"Stuart Rojstaczer."

"And you say that you're new here?"

"That's right. I've been here for a little over one week."

"Well, you know that this time of year, seeing that it's the beginning of the school year and all, everyone wants their phone installed."

"Yes, I know, but don't you have contingencies for that sort of thing?"

"Contingencies, sir?"

"Yes, you know. Hire some temporary extra staff to facilitate phone installation during the fall crunch. Have folks work overtime. Stuff like that."

"We're working as hard as we can, sir." And just like that I'm doomed. I've insulted the man by not being courteous and I've insulted his office by implying that they could do things more efficiently. What is worse, I don't have a clue that I've insulted him. It would take me years before I adjusted and learned to be polite enough to get favors done. I'm still learning and in retrospect, I would have benefited from being more courteous to others long before my move to Duke.

There were also issues independent of culture where I would have to make adjustments. When I arrived, the university was in the middle of a major transition in mission. Until the 1980s, Duke aspired to be a university with excellent undergraduate instruction and with a national reputation in its business, law, and medical schools. The only area where research was considered important was in the life sciences. Since the 1980s, it had been trying to become one of the handful of world-class research universities in the country. This was an ambitious goal for any time period, much less for the end of the Golden Age. World-class research universities had achieved their success in the United States on the backs of lavish federal funding. But the Golden Age of federal support was over. As a result, Duke did not quite have the infrastructure to facilitate such a change.

There were some significant shortcomings in infrastructure. The university computer and library systems were of marginal quality. Analytical chemical instrumentation in my department was inadequate (it has improved significantly since then). The research support office consisted of excellent personnel, but was woefully understaffed. As noted elsewhere, even the copy machines were at best dicey.

I knew that in comparison to many of my peers who had recently taken academic jobs, I had nothing to complain about. When I told friends and acquaintances with Ph.D.s where I had found a job, many were impressed and quite a few said that they were jealous. Sure, these friends and acquaintances had come from top-notch research universities, but there were precious few positions available in the top ten of the Golden Age (in no particular order): Harvard, Yale, Princeton, Columbia, Cornell, MIT, Michigan, Cal Tech, Stanford, and Berkeley. Most everyone who received a Ph.D. from these institutions had to find a position elsewhere. I was lucky. I had found myself a job at a university solidly in the second tier of American research universities. It was one of those thirty or so universities that laid claim to being among the top twenty universities in the country. In comparison, I heard tales of woe from peers who had taken positions at institutions where the resources for research were nonexistent and the talent of the undergraduate students was so poor that it defied description. At the end of the Golden Age, getting an academic position at any college or university was a bit of a coup.

Regardless of my relative outstanding fortune, getting my work done required extra effort compared to my colleagues at the best research universities. Consider the computer system. In 1990, most of the university's administrators did not have a computer or access to electronic mail nor did they want it. Because they were not, in general, computer literate, they saw no need to have a good computer system, and it showed. Computer connections to the outside world were made by slow speed modems, not the digital dedicated lines that had existed elsewhere (including at my government lab) for over three years. By 1995, my university had made dramatic improvements in its computer services, but in 1990 it was a computational backwater.

In my research in 1990, I was collecting lots of data in California and continually transferring data files from government computers to my computers at Duke. Once I had my phone, I could hook up to the government computers from my office. This in itself was a big improvement over my first seven weeks on the job. But computer connections to the outside world were poor and transferring files was taking me twenty times longer than I was used to. I felt like a commuter caught in a traffic jam. Something had to be done.

I did some investigating and found out that during remodeling, a dedicated digital computer line had been installed in the basement of my office building. It's just that no one was using it. After talking to a computer technician, I found that all I needed was about $500 in equipment and a few hundred feet of coaxial cable (the kind used for cable television). I bought the requisite materials and the computer technician and I ran the cable from the basement to my office. I was in seventh heaven. I had the same computer connections as the big boys at world-class institutions. It wasn't until fifteen months later that everyone else in the building was connected similarly.

There were other occasions where I had to work around the system to get things done properly. Somewhat like the computer connection in the basement, resources often existed, but they weren't advertised and you had to poke around to find them. Our library had a poor collection in the physical sciences, but the University of North Carolina was just twenty minutes away and it was easy to get a library card that allowed you complete access. The university did not have the instrumentation to perform the chemical analyses I needed either; but it was usually not too complicated to have government scientists or faculty members at other universities do the analyses for me in exchange for a small fee or coauthorship on research articles.

Even my problems with unreliable copy machines had a relatively easy solution. Initially, I tried using a nearby private copy shop for making multiple copies of grant proposals and research articles. It too had its problems. Then I used the university's copy services, but they were not conveniently located. Eventually they moved and were replaced by an X-rated bookstore, which caused an embarrassing situation for a new graduate student who I had sent out for some copying.

One day I surmised that if there was anywhere on campus that had a decent copy machine it must be in the president's office suite. So I went there and found, as expected, a top-of-the-line machine that produced peerless copies. The next time I needed to copy a proposal, I just walked into the president's office suite, acted like I belonged there, and started copying away. Several years later the university bought a few decent machines and put them in a cluster in a convenient location. Until then my graduate students and I used the president's copy machine for most of our proposals. Not too surprisingly, I wasn't the only one who had made such a discovery. Once, I had to wait in line while another young professor in another department used this copy machine for his proposal. He also had learned how to work his way around the system.

There was one situation, however, that I couldn't work my way around easily. University politics are hell. I've compared notes with friends in other departments and universities and have been somewhat heartened to find out that they can be much worse. I have only witnessed a little of the childishness, acrimony, and boorish behavior that exist in much greater quantity elsewhere. My departmental colleagues are generally honest, open, and fair. My university's senior administrators are less intrusive than average. The politics still have been hell. The amount of political intrigue at the government job that I held previously was barely visible in comparison to what exists at universities.

Henry Kissinger, who held more celebrated academic and government jobs than my own, once stated, "University politics are so vicious precisely because the stakes are so small." I agree that university politics can be vicious and also agree that the stakes are usually pitifully small, but I don't think that the two are related. We argue over small things because we are generally extremely detail oriented. As researchers and scholars, we typically derive great pleasure out of finding beauty in small things and the success of our careers depends upon us doing this well. For example, one of my best known colleagues at Duke had the talent to write an interesting and popular book on the unlikely to be captivating topic of the history of the pencil.

In our research we are all at least part-time miniature artists. The best of us have the ability to more than occasionally examine how each small piece we create fits into the larger picture. Even those who spend most of their time working on the broader aspects of their field usually understand the value of careful detailed work. This penchant for detail is great for research, but it is a disaster for politics. It means that we are incredibly skilled at making mountains out of molehills and spending meeting after meeting discussing issues related to the workings of a university that would put most anyone who isn't an academician instantly to sleep.

Then there is the issue of viciousness. My own department has seldom had vicious politics. We aren't all that nasty, but almost all of us possess a strange mixture of ambition, competitiveness, and social awkwardness. These are common traits in academia. University professors are often the types of people who like to work alone for many hours a day. Our success depends on a dogged determination to become one of the best in our field. We are given offices (within three days), furniture, telephones (within seven weeks), and are then left to our own devices to find a way to become successful. In our quest for achievement, we are more like independent contractors than employees. We are not, as a group, very capable of interacting with other people nor are we usually hired for our ability to work as part of a team.

Some may wish to ascribe lack of social skills strictly to those in the sciences, but from my experience it is clear that it is prevalent across the campus. Professors in the humanities simply dress a little better. The anger and resentment that sometimes suffuse academic politics usually stem from a combination of competitiveness and social ineptitude. It is odd that people who generally can communicate so well about their research are so incapable of group decision making.

With the end of the Golden Age, the political stakes aren't always small. Every department is fighting over very limited resources and the economic health of universities is being stressed. Even the elite universities with multibillion dollar endowments are working hard to stay afloat. Senior administrators are being forced to renege on prior commitments and are trying to find ways to cut their budgets. Professors and administrators became used to economic health and growth and most have never seen bad times. For many, the change has been hard to accept.

If Henry Kissinger was right, the presence of real stakes should have improved the state of university politics. Instead it has made them worse. As a result of the end of the Golden Age, political battles between administrators and faculty members have reached new levels of acrimony. For example, in 1993, Harvard's President Neil Rudenstine, who was accustomed to warm relations with faculty, was so shaken after clashing with faculty over cuts in faculty pension plans and health benefits that he took a leave of absence for two months.

I've never adapted to the supercharged nature of university politics. I don't think that I ever will. It's not that I haven't participated. I can slug it out with the best of them. I still loathe it. As Albert Einstein (whose son, Hans, was a hydrologist) once said about politics, it "makes the clean dirty and the dirty dangerous." I am tempted to relate some of my more lurid political encounters in juicy detail, but I will respect the privacy of my colleagues (and the wishes of my publisher's legal staff) and abstain. I will say that I can't understand how we can become so emotional over issues that are ultimately of little consequence. Lately, I've tried to adjust by avoiding faculty meetings unless they are absolutely critical. This change has helped me a great deal.

In moving into academia, I had to adjust to changes in my home life as well. At my previous job I had worked about fifty hours a week, a level of commitment to my work that allowed me a good deal of time to spend with my family. I enjoyed a close and loving relationship with both my wife and my daughter. I spent a lot of time traveling with them, doing chores around the house, and just plain goofing off.

But when I took my new job at Duke, I went into overdrive and worked seventy to eighty hour weeks. I was convinced that this was the only way I was going to be productive enough to get tenure. I liked the work and enjoyed seeing the rapid progress that I was making. Work can be quite addictive. Hard work is an activity that most everyone in society admires. Not only is there an internal sense of satisfaction at the end of the day, but your peers tend to smile when they notice that your office door is open almost all of the time.

Hard work has its benefits outside the office, as well. My colleagues and friends in hydrology are almost like a second family to me and in that family work reigns supreme. Once or twice a year I meet my fellow hydrologists at national meetings. The most common greeting I receive from them is, "Hello Stuart, what have you been up to?"

In response, I'm supposed to say something along the lines of "Man, I've been working hard," and then recite all of the progress I've made on my research projects.

All in all, working seventy to eighty hour weeks had its benefits. I had a sense of satisfaction and my peers both at Duke and at other universities were giving me pats on the back. There was only one problem. There is no way one can work this much and have any sort of reasonable family life.

My wife kept telling me that I was losing touch, but I didn't think so. I would still show up for dinner almost every day, notwithstanding that I usually showed up late and then returned to the office at night. I still spent half of Saturday and most of Sunday at home. I thought that I was having the best of both worlds, a productive career and a happy family life. Eventually, I managed to realize that all was not right with my lifestyle.

I became aware of how much my interactions with my family had changed one Sunday when for no particular reason I decided to watch an old family videotape. On the tape, a neighbor was interviewing my daughter, then three years old, and me. We sat on a couch and answered questions sometimes in unison and sometimes separately. In between questions, we would laugh and thumb wrestle.

I watched the videotape for about ten minutes. I thought about my relationship with my daughter now that she was seven. It was very apparent that after two and a half years of working long hours, I no longer shared anything close to the warmth and closeness I had with her before I took my job. It was almost certainly true that I no longer shared as much with my wife either. I felt awful. I hadn't been balancing both worlds at all.

I decided then and there that I had to cut back on my hours at work. If my productivity at fifty hours per week was not good enough to achieve tenure, then so be it. I know that there are many jobs that require long hours and know quite a few people who have worked at a hectic pace their entire adulthood. Some of them say that they have managed to maintain a reasonable family life. I don't believe them. I like having a strong connection to my family. I'm not willing to give it up for the sake of a job, no matter how satisfying that job may be.

15

Getting Tenure

During my first month on campus, I attended the annual meeting for untenured faculty sponsored by the university administration. The meeting consisted principally of a panel discussion about the tenure process. In order for us to get to know each other and the members of the panel, it was followed by a social hour. On the panel were members of the committee that made tenure decisions (the APT Committee, where the letters stand for Appointment, Promotion, and Tenure or "Almighty Police of Tenure" depending upon whether you have tenure) and some of the deans and senior university administrators. Every year we hire a few tens of new tenure-track faculty and it usually takes six years before you can be considered for tenure, so there is a potential audience of about 200 for these meetings. Of course, many of these people have already attended at least one meeting like this and don't feel compelled to come every year. Seated with me in the audience that year were about fifty untenured faculty members.

I looked at them, quiet, nervous, and attentive, wearing their name tags on their jacket lapels and dresses. Most of them looked younger than I, but not too much younger. The academic job market had become frighteningly competitive beginning in the 1980s. As a result, students took longer to finish their dissertations, and universities like mine had the liberty to hire faculty with considerable experience. The era of the twenty-something assistant professor (the title usually given to untenured professors) at universities across the country was mostly a thing of the past.

Some of them were dressed with an eye for fashion. Others had chosen to wear the uniform of the eastern academic (tweed and khaki for the men, dark, longish, knit dress with some tasteful jewelry and perhaps a scarf for the women). Others made no statement at all in their dress but like me simply looked unstained, unwrinkled, and recently bathed. In seven years time, less than fifteen of these fifty bright, young (by contemporary academic standards), ambitious faculty would still be at the university. A few, by a mixture of luck, hard work, and talent would become stars in their field and move to universities with loftier reputations. Others would move to universities of similar or lesser caliber for personal reasons (such as to be closer to a spouse with a job in another city). But most of them would leave before their tenure decision (because they became convinced they would not get tenure), or because of a negative tenure decision. Tenure is the one critical judgment by a university of your ability. Either you are promoted to a lifetime position or you are fired. It should be noted that university administrators don't use the word fired. Rather they use the phrase "let go," as in "Professor X was denied tenure and let go." It was no wonder that nervousness and anxiety filled the room.

The meeting began with addresses by some of the panel members about the nuts and bolts of the tenure process, including criteria for tenure. In discussing the role of teaching in the process one of the panel members (a senior administrator) said, "You are all likely good to excellent teachers, and your ability to teach well is a given. What will separate you is your research." Obviously, the statement about teaching was patently false. This obtuse way of communicating information is common of university administrators, and it is one reason why I never want a job in administration. I fear that I would be surrounded by people who talk like this or (worse yet) begin to talk like this myself. But I digress. Research was the key to success.

In the audience of untenured faculty was someone who I came to know fairly well. He was a nice guy, loyal to the institution, who loved students. They dropped into his office at all hours of the day seeking help, and in the following five years he won a few teaching awards. At the end of six years of service, he was denied tenure on the grounds of insufficient quality of scholarship and insufficient quantity of research dollars. In many ways it was sad to see him go, but by focusing so heavily on teaching, he chose to not play the game by the rules. He should have taken to heart what was being said at this meeting. The rules of the game, good or not, were being laid out in front of us.

The final and longest address came from one of the deans. He was a tall and rail-thin man with the mien and dress of the eastern urban academic. He waxed philosophical about the tenure process. I could feel my eyelids getting heavy, but I forced myself to listen. "The awarding of tenure is a lifetime commitment on the part of the university," he said. "We have to be absolutely sure that the faculty member will be an asset to the university. We cannot afford to award tenure to someone who has simply done well. We can only award tenure to those who clearly stand out. We may make a mistake in our judgment. However, it is important that when we make a mistake it is not one where we award tenure to someone who turns out to be of marginal intellectual value to the university. If we award tenure to such a person we will have to live with that mistake for decades. So when we err, we tend to err in not awarding tenure to those who subsequently may prove to be an asset to another university."

He continued in a similar vein. We had to stand out above the crowd of scholars in our field. Even if we did so, there was a chance that the university would make a mistake and deny us tenure. I didn't understand the purpose of what he was saying. Clearly, he wasn't trying to be inspirational. Perhaps he was simply trying to scare the hell out of us.

After this speech, it was time for an open question and answer period. But the final talk, as bleak as it was, left the crowd highly reluctant to ask any questions. Asking a question might be interpreted as a lack of self-confidence in your abilities or your status. And if you were labeled as unconfident, how could you stand out in a positive way? So first, there were a couple of basic questions from the crowd. When was the first progress review? The answer, during the fourth year. How does the first review differ from the tenure review? The answer, the tenure review requires letters from outside scholars and a careful evaluation by the APT committee.

Eventually, the anxiety of the faculty began to come out a little at a time. "What if you have a nonsupportive chair?" someone asked. The answer by one of the panel members was convoluted. Given the dark mood, I was tempted to interpret the lack of clarity in the answer as an indication that a nonsupportive chair was a definite problem. Increasingly, the session felt like group therapy. "My department doesn't seem to value the type of research that I do, even though they specifically hired me for my different approach to examining problems in my field," another stated. As far as I could tell, she wasn't asking a question, but was simply foretelling her own doom.

I was getting squeamish listening to this type of stuff. I am a scientist and like many scientists, I don't like to hear personal fears and anxieties expressed in public. All I wanted was to hear was the skinny on what it took to get tenure. I raised my hand from the back of the room and asked a question about grants that I discussed in an earlier chapter. Then I asked what I thought was a very basic question. "Look at us," I said. "We've got about fifty untenured faculty in this room. We were hired because of our academic promise, our potential to be stars in our fields. But the reality is that even though we may all be good and work like crazy, circumstances will only allow for less than 10 percent of us to be true all-stars. So obviously, unless you deny tenure to more than 90 percent of us, we don't have to be stars to get tenure. Just how good do you have to be at the end of six years?" This question elicited a big round of laughter from one of the panel members which traveled around the room. Even if I didn't get a straight answer, I thought that I had at least cut the level of tension. "Stuart," the panel member who instigated the laughter said, "at the end of six years you have to be a major league player in your field." A baseball metaphor, I thought. Here was a kindred spirit. No wonder he laughed at my comment.

"Well at this point," I said, "I would characterize myself as a utility infielder. I guess I have six years to get to be a starter, is that it?" There was a little more laughter.

"With that, I think it's time we adjourn for some drinks and hors d'oeuvre," said one of the panel members. And so we did.

In 1990, my university still thought it had a lot of money, and the food at the cocktail hour was very tempting. It was like being at a wedding. There was a mountain of shrimp on ice, hot hors d'oeuvre with crabmeat, prosciutto rolled around asparagus, filet mignon cut into small strips suitable for placing upon small toasts of French bread. On the large sweets table there were strawberries, small cheesecakes in foil, and (for a little southern touch) small slices of pecan pie. To drink, there was wine and champagne. I looked at this bounty and thought I should call my wife and tell her that there was no way I was going to be able to eat dinner. I compared this spread to a similar one my wife and I had experienced at the president's house earlier that year. I remembered the bottle of wine that I received in my mailbox on my first day with a note from my dean. We were definitely being welcomed graciously in terms of food and drink.

* * *

During my first six years, I attended almost every meeting of untenured faculty. It gave me an opportunity to see that year's lineup of the APT and my dean of the year (for some reason during my first six years, deans only lasted a year or two instead of the usual five to ten years). I was able to become a familiar face to everyone in charge of my tenure decision. I figured that it was easier to turn down an anonymous face for tenure than someone with whom you have broken bread. It should be noted that, in comparison to other untenured faculty, my near obsession with this meeting was unique.

The meetings followed the same format year after year, but in 1993 one aspect changed significantly. The university started to get money conscious and began to stretch its entertainment dollar. The contrast was startling, like that of a family that had lost its fortune due to a major dip in the stock market. The food at these meetings changed to miniature pigs in a blanket and some mediocre cheese and crackers. I no longer felt tempted to skip dinner. Other meetings of both faculty and alumni befell a similar fate. The end of the Golden Age meant an end to most of the opulence in entertaining.

The last time I attended this meeting, a senior administrator who I had come to know through the years came up to me at the skimpy buffet, smiled, and asked, "Haven't you heard this story enough?" He wore a gray suit that fit poorly. It was September and his skin was pale. It was clear that he had spent little time outside that summer.

I smiled back. "Yeah," I said. "I've heard it a lot of times. I just want to make sure the story doesn't change." I picked up a wine glass and took a sip.

"And has it changed?" he asked.

"Not much," I said.

After the first meeting of the untenured faculty, I thought about what I would have to do to get tenure. It's not that I mapped out an elaborate strategy, but I did not feel confident enough to simply "do my work and let the chips fall where they may." Like every untenured colleague that I had met, I felt a certain need to try and optimize my chances.

I knew I had to do two things over the six years. One was to impress my colleagues in my department. The tenure process consisted of two votes. The first round came from my department's tenured faculty. If this vote was positive, then my case would pass upstairs to the APT Committee for the second vote. I asked around and found out that the department vote had to be nearly unanimous for me to have a chance with the APT Committee. At the end of six years, the department had to be enthusiastically behind me or I would likely fail. Then I had to have a scholarly record that was also impressive enough for the APT Committee to vote for my tenure.

The ability to impress one's colleagues depends on a number of factors. One consists simply of local politics. At the end of six years, you cannot afford to have offended too many colleagues. In some university departments, this is nearly impossible to achieve. The problem is that there is so much political turmoil and factionalism that just the act of making friends with a colleague can create enemies. Departments like these tend to be graveyards for untenured faculty. I have friends across the country, excellent professors in fact, who were denied tenure simply because their departments were a political mess. Fortunately, my department was fairly free of this kind of political strife.

Some untenured faculty respond to the pressure to not offend by, like Jacob in the Bible, living a life of subservience for seven years. They do their best to avoid controversy, say little at faculty meetings, and willingly and enthusiastically do whatever the senior faculty asks of them. This approach can be successful, but it has a downside that goes beyond a loss of pride and self-esteem. After six years at a major research university, you are not only supposed not to offend, but in some way have to impress your faculty, and distinguish yourself from the crowd. It's hard to impress when you are serving the role of lackey. It's far better to try to achieve a balance between being accommodating and showing enough assertiveness that your senior colleagues do not view you as a rug.

Almost all untenured faculty that I've known have tried to achieve this type of balance, but if you are as assertive as I am, it's not possible. I am assertive to a fault. I can count the number of times in my life that I've tried to defer to someone on the basis of his or her seniority on one hand, and even then, I haven't shown much skill. (My wife says that my version of sucking up looks like most people's versions of simply being polite.) At the time of my tenure decision, I would not be able to rely on a record adorned with many instances of doing the right things for the right people. In retrospect, I was lucky to be in a department that did not require me to toe the line. As long as I did my job well, I would have the support of my colleagues.

Doing my job well meant getting grants, writing papers, and in the process developing a high degree of national visibility in my field. To a much smaller extent, it also meant carrying my fair share of departmental duties from teaching, to advising, to serving on departmental and university committees. Both my department and the APT Committee expected a high level of achievement.

The difference between my department's and the APT Committee's judgment of me was one of timing. First impressions were important with those I saw daily. Looking at tenure experiences of friends and acquaintances, it's clear to me that departmental judgments of junior faculty are almost fully formed within the first two years. I had to get off to a fast start and impress my colleagues from the beginning or it was going to be difficult to succeed. With the APT Committee, their evaluation was, by design, more distant and much less rapid. They would see my file twice. The first time would be during my fourth year review, but it was very rare to fail (and be fired) after only four years. The fourth year review was largely perfunctory. Their true evaluation would be based on my total accomplishments at the time of my tenure review. If I got off to a fast start, but did little in subsequent years, my colleagues might still be inclined to approve me for tenure. But the APT Committee wanted to see evidence of sustained and significant progress in scholarship and university service every year.

Because the APT Committee was composed of people well outside my area of study, they had to rely on outside evaluations of my scholarship. Some of them came from within my department, but the most critical evaluations came from those in my field outside the university. During the year of my tenure decision, twelve well-known leaders in hydrology at other universities would be contacted and asked to write a critique of my scholarship. They would receive copies of what I considered to be my most important research articles and a copy of a resume with a list of my scholarly accomplishments and a record of university service. These faculty were required to be at "arms length," which meant that friends or people with whom I had worked were disqualified from participation.

The twelve letters from outside faculty had to be positive in tone. Most important, all of these critiques were required to include an answer to the following question: "If Professor X were in your university would he/she receive tenure?" A single evaluation that answered no to this question could result in a negative tenure decision. Two negative answers meant almost certain denial of tenure. By the time of my tenure decision, I had to develop enough national visibility in my field to assure that these critiques would all be favorable and, hopefully, that most would be overflowing with praise. They had to indicate that, within my field, I was a major league player.

Tenure is more or less the equivalent of achieving partner in an accounting or law firm, and just like most of those trying to achieve partnership, I worked like hell trying to achieve tenure. I worked evenings and weekends on a regular basis for the first two and a half years. After that I slowed down for family reasons, and replaced many of the evenings and weekends with the occasional all-nighter. (I still pull occasional all-nighters. Once, when a few students noticed me doing this, they commented that it was nice to see that even some professors leave things until the last minute.)

It's a phenomenon of the end of the Golden Age and the tightening of the academic job market that the standards for tenure have risen markedly. Because far more people want academic jobs than there are positions available, universities can afford to be very picky in whom they chose to promote with tenure. At the most lofty universities, the standards have generally become ludicrous. The expectations are that at the end of six years you are more than just a major league player. You have to be an all star in your field and less than ten percent of assistant professors are given tenure. As a graduate student at Stanford University, I noticed the preponderance of young (under forty) professors who were either never married or divorced. When I asked a senior faculty member at Stanford about this he said, "The truth is that if I had to do then what these junior faculty have to do now to get tenure, I would never have become a university professor." Like him, I am not willing to pay such a cost to my personal life, although the prestige of being at a world-class research university will likely always be alluring to me.

Although they were high in the early 1990s, the expectations at Duke were to my mind reasonable. (They may, however, eventually rise to virtually unreachable levels. One way to undermine tenure is to make it nearly impossible to get.) Mostly, I had to make sure that, at the time of my tenure decision, the leaders in my field knew and respected my scholarship. As noted in another chapter, I spent a good deal of time writing grant proposals and was fortunate enough to have considerable success in getting grants. But scholarly visibility does not come from writing grants. It comes from performing research, presenting research results at national meetings, and writing research results into scientific papers. I spent most of my time engaged in these activities. I enjoyed this aspect of my work a great deal. While the pressure of trying to get tenure forced me to write many of my grant proposals, almost all of my energy toward research came out of an intrinsic attraction to solving interesting problems.

I knew that my chances for tenure would be enhanced if I won an award for my research. Neither my departmental colleagues nor the APT Committee knew the field of hydrology very well, so their ability to evaluate critically the quality of my research was limited. Aside from external letters of evaluation, their judgment of my research was highly dependent upon largely symbolic indications, like awards, that my research was having significant impact.

In general, one cannot apply for an award. You have to hope that your research excites your peers enough that they choose to bestow you with special honors. In my field, awards are generally reserved for those who have had a long and distinguished career in research. However, with the end of the Golden Age and the tightening of money and opportunities, there has been an effort to create financial awards strictly for new faculty. The idea behind these awards is give a select number of new faculty some of the resources they need to get off to a good start in their research careers in what are generally tough times for getting grants.

The National Science Foundation (NSF) is the most prominent organization that provides this kind of prize. It began to offer its "Young Investigator" awards (most recently called CAREER awards) in the early 1980s. Typically, these awards offer up to $100,000 per year for five years to new faculty (faculty who have earned their Ph.D. in the last five years). What is most wonderful about these awards is that the money can be used for any research or teaching purpose. As might be expected, Young Investigator awards are hard to get. In all of the earth sciences (the combined fields of hydrology, oceanography, geology, and geophysics), about six of these prizes are given per year. Unlike any other kind of award I know, you don't have to be nominated. Instead you fill out an application.

At the time I was eligible, the application for this award was about five pages long. It consisted of a two-page resume including your record of publications, and one-page statements of your past research and teaching accomplishments, future research plans, and future teaching plans. Also necessary was a letter of recommendation from your department chair, and three letters of recommendation from anyone of your choice. It was expected that these letters would come from senior world-class scientists and that they would contain glowing assessments of the applicant's past success and future promise. Applications were evaluated by panels of scientists that represented broad areas of research. In my case, the panel consisted of a wide mix of earth scientists, mostly from universities.

In their search for research money (and prestige), virtually every faculty member at a research university who is eligible for a NSF Young Investigator award applies. Except for my first year, I applied every year that I was eligible. Most awards are given to those in their last year of eligibility so I wasn't at all surprised to receive rejection letters from my first two applications. With each rejection letter came written evaluations of my application from the evaluation panel. While the panel changed every year, and the likes and dislikes of a given panel are always somewhat unique, these evaluations were extremely helpful. They helped me improve future applications.

In the first two applications, I stressed the practical value of my research. I had thought that this was important, especially since the instructions explicitly asked you to explain the value of your research to society. However, in their written evaluations it was clear that the panel did not think all that highly of my research. My application led both panels to believe that my research was rather pedestrian and unexciting. Also, I noted that the panels tended to give awards to those who had managed to get a research article or two published in *Science* or *Nature*, the most prestigious journals in the scientific community. These journals publish articles that have broad interest, rather than articles of interest only to specialists in a given field. Hydrologists hardly ever publish articles in these journals.

On my final application, I paid serious attention to the previous evaluations. I completely rewrote the descriptions of my research. I made the assumption that the societal value of my research was so obvious that it didn't even need to be mentioned. What needed to be stressed was the fundamental, cutting-edge nature of the research and its inherent intellectual value. I wrote the application as if it was an essay championing the virtues of hydrology (and in particular my own research) to the scientific community. It didn't hurt that by then I had already published one article in *Science* and had another one in *Nature* in press.

It was a somewhat unfortunate coincidence that my final year of eligibility was also one when my university decided that it needed to apply quality control to the award applications. Previously, these applications had received the required signature from the Office of Research Support pro forma. That year, however, a change was made. Every application for a Young Investigator award was read through carefully and was returned for corrections. When I sent my application to be approved for submission, it was bounced on the grounds that I hadn't included any information on the societal relevance of my research.

I went to the Office of Research Support and tried to explain why I thought my approach in writing this application form was correct. "I know this community," I said. "They already know my work is societally relevant. What they don't know is that my work is scientifically valuable, as well." But the office was adamant: They would not approve my application unless I included a section on the value of my research to society. Since the application was already as long as it could possibly be, this meant that I would have to remove some material to make room for this change.

I was in a quandary. I knew that the administration felt it was trying to help me, but if I made the demanded changes, I would hurt my chances of receiving this award during my final year of eligibility. If I didn't make the changes, I could not apply. In the end, I decided to use some sleight of hand. I submitted an application with the requested changes for internal approval. But the application I actually sent to NSF was the one that I had originally written.

When I won the award, I received congratulations from the senior administration of my university and an article about me appeared in the campus newspaper. My daughter, who shares more than a little of my critical sense, said to me, "But dad, you aren't young. How can you win a Young Investigator Award?" I was thirty-eight years old at the time and my daughter was ten going on twenty-seven. In response to her comment my family refers to this prize as a "Middle-aged Investigator Award."

Tenure evaluations usually take place at the end of six years. At my request, my evaluation took place at the end of my fifth year. There were some political reasons for this slight acceleration in the process. I had been hired into the university's Department of Geology in the School of Arts and Sciences. Due to low undergraduate student enrollments and the financial pressures of the end of the Golden Age, this department was slated to be dissolved during my sixth year. Our faculty were to be moved and reorganized into a Division of Earth and Ocean Sciences in the university's new School of the Environment. Being evaluated a year ahead of schedule would avoid any complicating problems that might come with the reorganization.

By the end of five years, there was also a pervasive sense among the tenure-track faculty in my department that I was ready to be promoted. I had done most everything expected of me and I now had a prestigious award.

I put together my tenure file. It consisted of my resume and several statements on my research and teaching accomplishments. There was little, if any, drama associated with the tenure decision. My department voted unanimously in favor of tenure, and in December submitted their evaluation to the APT Committee for consideration. In early spring, my chair was required to go to the APT Committee to discuss my tenure case. At that meeting, my chair was strongly supportive of me receiving tenure. A few weeks later, I received a call from my departmental chair congratulating me on being promoted to associate professor with tenure. Then came a formal letter from my dean. I was now a fully accepted member of my university's community. I was granted the privilege of having a position as professor for as long as I cared to hold it.

Sure I had worked hard, but like all happy stories, there was more than a little luck involved. I've seen too many people deserving of tenure not make it. For example, every career has its peaks and valleys and some senior faculty members, when they are solicited for an external letter of evaluation, have a warped predilection for ignoring achievements and focusing on the negative. If a university knowingly or unknowingly solicits letters from such professors, the result can be disastrous for the tenure candidate. I managed to avoid this land mine. I was also lucky enough to be in a department that come tenure time almost always put aside personal differences and strongly supported their junior faculty.

Over the past decade, there has been a public debate about the worth of tenure in universities. Arguments have been put forth that tenure is an anachronism. Lifetime commitments to employees are no longer expected in the corporate world, and many see no reason why they should be in university employment. Tenure is also criticized because it kills the possibility of accountability for professors.

Imbedded in these arguments is the assumption that tenure allows for the accumulation of deadwood in the professorate. Somehow, because administrators cannot fire faculty, universities must contain a vast sea of professors who are accountable to no one and do little work. Also, faculty are viewed as an impediment to necessary structural change in the university. The elimination of tenure would allow for a supposedly justifiable shift in the power structure of the university toward the administration.

It is true that because of tenure, we probably have more deadwood than most corporations. Many departments can identify one or two faculty members who are more or less a waste of space. They do not teach well and no longer perform any significant research. However, my own observations suggest that they represent a small percentage of professors. The truth is that we already have a system in place to assure some level of accountability. In the first place, we have a formal annual review. Every year, I have to prepare for the department chair a document that details my teaching, research, and university activities for the previous twelve months. I cannot be fired, but the chair's decision on whether I receive a raise in salary and how much that raise will be is directly tied to the level of productivity that I have shown over the previous year.

In the second place, there is strong peer pressure for faculty to be productive. A professor who is frequently absent from his office and does little research will likely be ostracized by faculty in his department. Yes, it's very possible for such a person to have a thick skin and ignore whispers in the hall and terse exchanges between himself and his colleagues. But for a few, fear of disapproval by colleagues is a valuable motivator.

I know that the public has an image that professors are lazy, but personally I don't find laziness to be all that common. I admit that professors, including myself, do not teach as much as we should. However, the current system demands that we spend most of our time doing research rather than teaching. Almost everyone I know is working hard. It's just that research is a much less visible means of working than going out in front of a class and lecturing.

Without tenure, we could eliminate some of our deadwood. It is also true that the elimination of tenure would allow a streamlining of the university decision-making process. Because faculty would be more strongly beholden to the university administration for their jobs, they would have a much smaller say in how the university is run. Decision making would thus be more centralized. Universities would be free to follow a management model more close to that of corporations. There would be little buffering of decisions made by university presidents, provosts, and deans.

Wholesale transplanting of corporate organizational models to universities is, however, not a good idea. Universities differ from corporations in many ways. One key way is that corporations have many employees who aspire to positions of management. It is more or less possible to find corporate leadership by culling from the best and brightest in the employment pool. Yet this is not possible in universities. Almost all or our best and brightest would much rather remain professors. They have no wish to move into administration. When someone takes an administrative position, I feel pity for that person and I am not alone in this emotion. Universities have a very small pool of faculty interested in being a dean or president. With the end of the Golden Age that pool has become even smaller. It is not particularly appealing to be in charge under a backdrop of stressed resources.

University administrators across the country share many more personality traits with politicians than with corporate executives. As a group, they are a strange mixture of those who have chosen administration out of a sense of duty to the institution and those driven by a personal need to be in charge. Some are honest and some are corrupt. Some have strong natural management skills and others do not. There needs to be a mechanism in place to assure accountability from university administrators.

Some level of accountability is demanded by a university's board of trustees. But trustee boards, which consist largely of wealthy alumni, are not aware of the day-to-day workings of a university. Faculty have an inside knowledge of how universities work and have a vested interest in making them work well. With tenure, they can provide an essential check on university administration.

Under a tenure-based system, the balance of power belongs to university administration, but faculty have a significant say in the running of the university. Faculty don't decide how new appointments are distributed, but they are in charge of choosing who gets hired. They are in charge of the tenure process. They are nominally in charge of curriculum (faculty committees decide curriculum issues, but appointments on these committees are made by the university administration, and they often stack committees with faculty members whose opinions mimic their own). But almost all issues related to resources such as faculty salary raises, office space, and money for research equipment and computers are made solely by university administrators. Because of this, even with tenure, there are strong financial and programmatic incentives for faculty to cooperate with university administration. And they usually do.

Many university presidents have lamented that our current system of decision making gives faculty a significant ability to obstruct change. The fact is that in very few instances does this occur. It is almost always possible for the university administration to get what it wants by being patient and flexing its muscle. For example, when my university administration first proposed dissolving the geology department and moving its faculty to a different school, the faculty almost unanimously rejected the idea. But after two years of administrative pressure and some modest financial concessions, they voted overwhelmingly in favor of dissolving and moving. Faculty provide a valuable filter to change. Sometimes university administrators come up with truly bonehead ideas and some of those ideas don't make it through this filter.

When, on occasion, faculty make the effort to obstruct proposed changes, it often is justified. For example, two decades ago, our administration pushed to house the Nixon Presidential Library (Richard

Nixon was a Duke alumnus). It beats me why any university would want to be strongly associated with the only U.S. president forced to resign from office, and this idea was bounced by the faculty.

Even without the elimination of tenure, it should be noted that over the last two decades, tenure has been undermined by university hiring practices. Universities across the nation have frequently replaced retiring tenured faculty with part-time and full-time contract staff ineligible for tenure. This trend has been so pervasive that the number of professors on contract appointments in the United States now exceeds those who are tenured or in a tenure-track appointment. At my university, we have increasingly hired nontenure-track faculty and given them the title of "Professors in the Practice." Tenure may not be formally eliminated at many institutions. But by hiring contract staff and raising standards for tenure, universities are making it more of a rarity. This gradual change, if it continues, will cause a loss in the already tenuous ability of faculty to influence university decision making.

If tenure is eliminated (and I anticipate that it will eventually be eliminated at many institutions, particularly at public universities), any gains in efficiency will be counterbalanced by the loss of a valuable political resource. Tenure allows faculty a small, but meaningful, role in university governance. It also allows faculty to publicly criticize universities and to suggest where changes could be made regardless of whether these changes would be popular with university administration. As a modest example, I would like to point out that, for better or worse, the book you are reading would never have been written by the author if he did not have tenure.

16

Rolling the Dice

Every once in a while, I fall prey to "End of the Golden Age Syndrome." I look at my older colleagues and am filled with a mixture of envy and anger. I'm envious because I think that they had it so good. They had ample money for research and were greatly respected by the public at large. I get angry because I think that they are largely culpable for the relative lack of respect universities currently face in society. I have a laundry list of complaints about my older colleagues, both professors and administrators. They had a tendency to be inattentive to the educational needs of undergraduates. They seemed unable to comprehend just how lucky they were and became mired in petty politics. They decided to use the university as a tool for social political advocacy. They became so enamored of growth that they didn't stop to examine just how bloated and expensive the enterprise had become. They developed an unwarranted sense of entitlement. If they would have been more responsible, universities would not be in such a mess today. Most of the time I think that blaming my colleagues for much of what is wrong is ridiculous. Whatever degree of truth is contained in such complaints, it is a waste of time to think about them. Sometimes, however, my emotions overflow.

In March of my seventh year I was in such a mood and I couldn't shake it. Normally, I'm quite happy in the spring. It's the best time of year in this part of the country. That year, however, I wasn't buoyed by the change in weather. I was dissatisfied with being a professor. A faculty election in my school was nullified by the administration because they didn't like the way it had turned out. It was increasingly becoming clear that this maneuver was part of a trend of ignoring faculty opinion in decision making. I was still getting significant amounts of grant money, but it was tougher than ever to find funding, and the amount of money assigned to every new grant was shrinking. In terms of student response, my classes were going well, but I felt that I had succumbed to external pressure and stopped fully challenging my students. I started to question whether I was earning an honest living.

But then April came, and because of a few events at the university, my mood improved considerably. (Sure, T. S. Eliot thought that April was the cruelest month, but I never understood him anyway. How could anyone leave a perfectly respectable baseball town, St. Louis, for London?) I started feeling much better about the state of universities in general and my own in particular. Times were tougher than they once were, but they still were pretty good. Universities were subject to much just criticism, but the public still valued what universities had to offer. The old guys in the university (there were hardly any old gals) had done the best job they could during the Golden Age, and it was not clear that we young faculty would have had the prescience to make appropriate adjustments had we been in charge. Despite the significant imperfections of the job, I was damn lucky to be a professor.

My mood started to change shortly after April Fools' Day. First, I attended the dissertation defense of my first Ph.D. student. Dissertation defenses are usually happy occasions. The student has essentially completed writing his dissertation, and presents before a public audience a twenty-minute synopsis of the results. If the candidate has performed well, then the question and answer session that follows is often a joyous event where the student shows off his or her expertise and talent. My student had done his job well. It was a pleasure to watch him at his defense. He was at ease and he conveyed his confidence to the audience. I was proud of him, and proud of the whole process.

His research was related to the movement of contamination in groundwater. The work had direct application to cleaning up contaminated groundwater and predicting susceptibility of water supplies to contamination. I had never thought that I would work on such research. However, when he came to study with me, he convinced me that this would be an interesting and fruitful research area to pursue.

He was right and I learned a lot from him along the way. Over the years, he had developed skills that would serve him well in life. There were still quite a few academic job openings in hydrology, and he had a very good chance of finding a professorship. When I shook his hand to congratulate him on passing his defense with flying colors, we were both grinning.

Then toward the end of the month I had a visit from a prospective undergraduate student and her mother. They were from New Hampshire, and had come to visit prior to making a final decision on where the young woman would go to school. She had been accepted for admission both at Stanford and my university. She wanted to talk to someone who had been at both places and had received my name from a Stanford faculty member. There was a fairly strong family resemblance between the mother and daughter, but their demeanors were very different. The mother, her red hair cut short, was about forty-five years old and very serious. I could tell that she was a good listener (listening is, admittedly, not one of my better skills). Her daughter was a little shorter and wore her red hair long. She was somewhat laconic, but exuded confidence and good cheer.

They were being very careful in making their decision. Choosing where to attend college was for them a critical issue with significant future consequences. Early in my career, I didn't understand why parents and prospective students sweated over school choices. But over the years, I think I've begun to understand what they are feeling. Every school has its own personality, strengths, and weaknesses. Like finding the right job, students want to find the university that best fits their needs. Given the cost of college education, they want to make sure that they are spending their money wisely.

Most prospective students, if they are accepted to both Stanford and Duke, will go to Stanford. Stanford is more prestigious (*U.S. News and World Report* yearly ratings of universities notwithstanding). But this prospective student was unusual. "I think I like this place more," she said.

"What do you see the differences are?" I asked.

"Stanford has the reputation, but it seems friendlier here," she said.

There was probably some truth to this statement. Our students generally are outgoing. Stanford tends to attract students who are a little more introspective and get a kick out of playing with their computers. I asked her about her interests. She had worked on quite a few local environmental problems in her small town. She had even organized her own grassroots group to create more parks. These were good people, I thought. The girl, if she kept pursuing her convictions, had a bright future ahead of her. Both Stanford and my university should be happy to have her as a student.

They asked me what I thought the differences were between Stanford and Duke. I didn't feel any need to sell them my university. Both are good schools, I told them. Each has its advantages. "Stanford is hands down a better research institution," I said. "If you plan on getting a Ph.D., there might be certain advantages in going to Stanford as an undergraduate. There might be a greater potential to work on world-class research projects. We also perform significant research, but pay a little more attention to undergraduates in the classroom." I asked the daughter a few questions about her goals. It was not clear whether she wanted to go on for a Ph.D. We talked for a while longer. All along, they seemed surprised that a professor would take the time to talk with them for an hour or so. But I didn't think that this was odd. As I said good-bye and shook hands with them, I wondered if our conversation would significantly influence their final decision. It was easy to get distracted by the problems with today's universities. But the mother and daughter in front of me saw a great deal of value in undergraduate education, and there were many like them.

The following day something happened that reminded me of a time when I was in high spirits. I was walking along the university's main quadrangle and spotted a student approaching me from the opposite direction. He was wearing an engineering school T-shirt and looked vaguely familiar. He was tall with blond curly hair and walked swiftly and proudly. I thought that maybe he had been in one of my previous classes, but I couldn't quite place his face. As he approached closer, however, I suddenly recognized who he was. My mind flashed back to the impromptu campus tour that I gave to a family four years before (described in the opening chapter). He was the high school kid in that family. He looked strong and athletic (being twenty years old has its benefits). His posture had improved, but otherwise he looked the same. Perhaps three years at Duke had helped to eliminate his former habit of slumping or maybe it was just growing up and being away from home.

I laughed to myself remembering my conversation with his Purdue-educated dad. I had been dead wrong in my prediction of where this student would go to school. He had been accepted and chosen to come to Duke. In about a year he would graduate. Perhaps the tour that I gave his family four years before was as awkward for him as it was for me. But it must have not hurt him that much. In the end, his father decided that my university, expensive though it may be, wasn't so bad after all.

By the time graduation and commencement came in mid-May, I was feeling much better. My case of the blues had receded. I am not a fan of pomp and circumstance, but I love graduations the same way that I love weddings. They are such inherent displays of optimism and idealism that the positive mood they create is infectious. I always try to attend some aspect of graduation week. The students are relaxed and tend to be more talkative than they are during the school year. I especially like talking to the proud parents. However, until my seventh year I'd always avoided commencement. It was just too big and impersonal. That year my first Ph.D. student was graduating and I felt like I owed it to both him and his family to attend. Former President Jimmy Carter was giving the commencement address, and it is not often that you get to hear a former president give a speech. So I ordered a cap and gown from the university bookstore (they kindly provide free rental of caps and gowns to faculty) and took part in the procession.

It was a typical hot and sunny day. I was sitting in a folding chair on the grass field of the football stadium. (On account of our lack of football prowess, our stadium has been called the prettiest place to watch the ugliest football in America.) The family and friends of the graduates sat in the stands. Many were using makeshift fans and others shaded themselves with umbrellas. The graduating students also sat in folding chairs wearing their caps and gowns. My seat was close to a block of seniors getting their B.A. degree. The students had a happy glow about them. The overall mood was decidedly upbeat.

Technically, I was supposed to take off my cap during the ceremony. But I had gone to a barber the day before and for some reason he decided without any consultation to give me a crew cut (I could not see what he had done until he gave me back my glasses). I knew that my scalp would burn if I took off my cap. I partly listened to the ceremony and partly observed the crowd. I noticed two seniors in the front row of the student section who were obviously boyfriend and girlfriend. One of them, I recognized as the daughter of a faculty member. They were a cute couple and they made me think of the time when my wife and I were undergraduates.

There were many speeches about the meaning of education from those representing the student body, faculty, and university administration. If there is ever a time when we talk in lofty tones about the noble educational objectives of a university, it is during commence-ment. Normally such oratory seems foolish, but on commencement day it works perfectly. As I listened, I thought about my own views on higher education. They were as idealistic and ambitious as those that I heard coming from the podium.

For me, higher education is principally about teaching students to think well independently and exposing students to the best intellectual achievements of civilization. It allows students to gain an appreciation of the world in which they live and of what humans can accomplish. Over their time at the university, I want them to develop the skills to learn on their own and to understand that there is always value in learning more. My view is that people with greater intellectual aware-ness have an ability to fully enjoy the world around them and have a greater positive impact on society.

This view is an old one and it is one that I share with many in education, past and present. When someone tells me that they are in school principally for the credentials to get a good job or to get good grades to go on to professional school (this happens to me more than occasionally), I almost wince. Sure I know that it is important to have pragmatic, practical goals, but higher education should have a life-enhancing component that goes beyond finding the skills to put money in the bank. When we are doing our job correctly, students gain a broad appreciation of the value of thinking well.

The topic turned away from education and toward a more universal theme when President Carter addressed the crowd. He talked about reaching out to those who were less fortunate. It was an appropriate speech for a Sunday morning commencement, and it was the best talk I have ever heard him give. Listening to Carter, I thought that he had missed his calling. He was not successful as a president, but he might have made an outstanding minister. His spiritual beliefs and concerns for creating a moral society undoubtedly contribute to the successes in his life that have followed his presidency.

After commencement, I went with my wife to the home of my graduating Ph.D. student for a reception. Both he and his wife were from Virginia and relatives from both sides of the family had come to attend commencement. It was an afternoon filled with speeches, toasts, and warm feelings. The families had a long history of attending uni-versities. Getting a Ph.D. was a milestone that made my student's relatives proud. I thought of my own Ph.D. graduation ceremony. I had skipped my B.S. ceremony (such ceremonies were not particularly fashionable at the time), but my mother was not going to let me deny her the pleasure of seeing me get my Ph.D. diploma. I had attended commencement thinking that I was doing it strictly for my mother, but sometime in the middle of the event, I felt that I was getting something from the experience as well.

By the third week of August 1997, my spring bout with End of the Golden Age Syndrome was a distant memory. I was in the local airport with my family. We were flying to Italy where we would be spending the semester overseeing a junior year abroad program sponsored by the university. I had taken a lot of good-natured ribbing from my colleagues about the "hardship" of this duty. The university had wanted a science professor to oversee the program and teach classes, and I had been the only one in my entire university willing to take the job. True, a few had wanted to go, but had found it impossible to find a way to have their spouse come along. Others had felt that four months overseas would be disruptive to their children's education. I was lucky because my wife was not tied to a job. We thought that living in Italy for four months would be a valuable experience for our daughter and all of us looked at this journey as a wonderful opportunity.

We walked through the baggage claim section on our way to the ticket counter. I noticed the crowds of students waiting for their bags. They had arrived to start the school year. I looked around to see if I recognized any of them, but no one looked familiar. I thought about August of 1990, when I had arrived in this same airport on my way to my new home and job. Seven years had passed and all the students that I had seen then had long since graduated.

There had been many changes in the world since that time. Domestically, it looked as if we would be able to balance our nation's budget for the first time in thirty years. As for politics abroad, the bloodshed in Bosnia had come and gone, although there was no real resolution in sight. On the baseball scene, my Oakland A's had a new coach and virtually all new players. I found it hard to keep track of major league rosters because they changed as swiftly as the weather.

I probably looked old to the students in the airport. My daughter, who walked alongside me, would be going to college in five years. I was as old as many of the parents of these students. Seven years before, they had looked too young to be university students. By 1997, I was used to being around eighteen to twenty-five year olds and the fresh tenor of their voices. They had been my audience in classes for a long time, and whatever questions I originally had about teaching them had long since been answered. Over time, I had learned to establish a solid rapport with students. Sometimes, I heard faculty complain about how students were better back in the "good old days." I disagreed. The main problem wasn't with our students. They could only be expected to work and think as hard as our current academic standards demanded.

As I stood waiting at the ticket counter, I thought about my university experiences over the last seven years. On the negative side, I had learned more about university politics than I had ever cared to know. On the plus side, I had learned how to teach well and had broadened my research so that my work was of interest to a wider audience. Had my university changed much over the intervening seven years? In some ways it had changed. It was a little more concerned about undergraduate education. For example, it had developed a new program so that freshmen would have greater access to small size classes. Our main focus, however, had remained the same. We kept running the university as if the Golden Age was still in existence and growth in research and infrastructure were a necessary component to success. Over the last few years, we had developed an awareness that there were financial limitations to growth. But we still wanted to grow and kept hoping that renewed prosperity was just around the corner.

I looked at the students walking in the airport terminal. After seven years, I had a pretty good understanding of what they could expect in their year ahead. Given the lengthy period of the Golden Age, it was not surprising that the education that these students were about to receive was not a whole lot different than the education that I had received. It was both easier and more personal, but the curriculum was much the same. We had changed slowly with the times. It was likely that changes in the future would be much more rapid. I couldn't help but believe that universities would be forced to be more attuned to the needs of society.

I tried to imagine what higher education would be like for the next generation of students. There were some who predicted a bleak future for higher education and the American research university. They said private universities like mine would only be affordable to the wealthy. State universities would either degrade in quality due to lack of state support or would decide to raise tuition to levels that would shut out those who weren't children of the upper middle class. Students would show their displeasure with traditional forms of liberal arts education by avoiding areas of study that were not directly tied to job availability. Distance learning afforded by better telecommunications and computers would improve public access to higher education and conversely cause a dramatic decrease in the number of professors needed.

Research money would increasingly be directed toward projects that gave faculty members little freedom to publish their results or to explore innovations in-depth. To a large extent, all of these bleak predictions were simply extrapolations of trends.

I didn't know to what extent these trends would continue. I didn't possess any supernatural ability to predict the future. Prediction is always a dicey business, but I hazarded a guess that there would continue to be high demand for our research and education. We would continue to award undergraduate, graduate, and professional degrees. But what those degrees represented in twenty-five years (when the children of these students would be in college) would likely be something quite different. In many ways, I had liked the model of the Golden Age, and I wished that it could continue. But I knew that we had held onto this model as a way of doing business for far too long.

I thought about what would happen if we decided to perpetuate the status quo. Following our Golden Age model, we would continue to strive for growth in faculty size, research dollars, and infrastructure. This strategy would have its advantages. The additional financial pressures it would produce would be partially counterbalanced by the sense of well being that comes from being part of a growing enterprise. It's almost always uplifting to see new buildings under construction and additional faculty in departments. Although federal funding would continue to favor safe and relatively uninteresting research projects, path-breaking work would still be done and our country's status as a leader in university-based research would be maintained. Our lack of institutional effort in creating inspired undergraduate education that stretches the best and brightest would be mitigated in a piecemeal fashion by the many professors who care about teaching undergraduates. Most of our efforts for undergraduates would be oriented toward providing a pleasant setting to spend four years and, for the elite, private schools, the trappings of prestige at an increasingly daunting price. Higher education at elite universities would eventually become (if it was not already) the equivalent of a hand-sewn suit, a quality product for the wealthy and discriminating.

Personally, I didn't want to be part of such a university, and I doubted that this was the direction that we would take. For almost all universities, continued rapid increases in tuition would not likely be viable. Over the next twenty-five years, my university would likely become leaner (it may become so lean that I may be forced to retire early). In response to any future changes, I might develop a longing for my early years as a professor. I might, like my older colleagues, be thankful that I experienced some of the bounty of the Golden Age.

I didn't think that this would happen. I could well imagine that over the next twenty-five years I would have less time for research and greater teaching responsibilities. Neither of these prospects scared me. I was looking forward to a change that moved the American research university in a direction away from nostalgia. The Golden Age was gone for good. Perhaps we would recognize that future growth was neither necessary nor desirable.